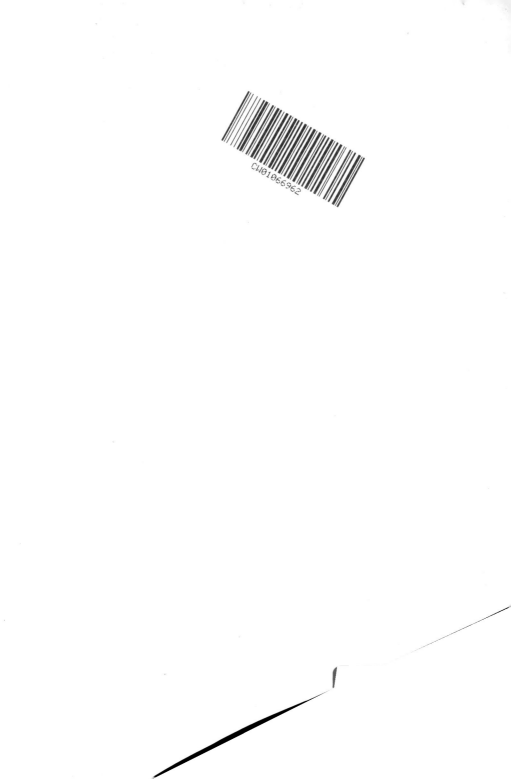

*Introduction*

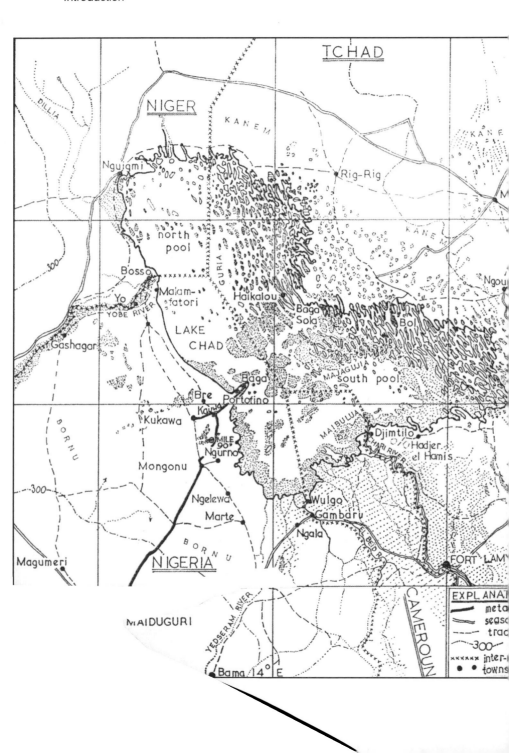

*Lake Chad versus The Sahara Desert*

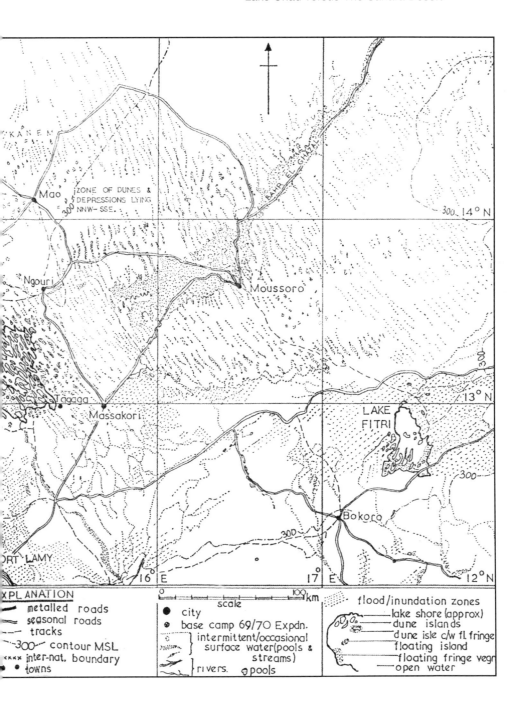

Sylvia K. Sikes

# LAKE CHAD VERSUS THE SAHARA DESERT:
a great African lake in crisis

Foreword by
SIR VIVIAN FUCHS

By the same author:
The Natural History of the African
      Elephant (1971)
Lake Chad (1972)

Published by
MIRAGE, NEWBURY

This book is dedicated to the memory
of

# Erika von Bernuth

*First published 1972*
*©1972 Sylvia K. Sikes*
*Printed in Great Britain for*
*Eyre Methuen Ltd,*
*11 New Fetter Lane, EC4P 4EE*
*by Western Printing Services Ltd*
*Bristol*

*ISBN 413 27590 6*

*Revised Edition published 2003*
*© 2003 Sylvia K. Sikes*
*Printed in Great Britain*
*for Mirage Newbury,*
P.O. BOX 6025
NEWBURY, RG14 5XS, UK

*by*
*Pentalith Ltd*
*Craven House*
*Craven Road, Newbury RG14 5NE*

*ISBN 0-9544079-0-3*

# PROLOGUE
(As if written at some time, perhaps within the next 100, or maybe the next 1000 years)


I knew Lake Chad aglow and shimmering in the heat,
A giant puddle in the desert haze
With birds and beasts, boats, fish and fishermen;
With floating islands, reeds and hidden dunes.
I sailed her subtle waters both in storm and calm
And ever learned to love her fickle moods.

I should revisit Chad today if she were real,
If yet her sedges sighed and waters rose and fell . . .
But Chad is just a mirage now,
A jest of nature to belie man's huge conceit
In thinking he alone could run God's world
With new techniques computed by machine.

Thus man has tried to harness Chad.
Alas! the lake dried up and left a sandy desert in its place
With ancient artefacts of Space-Age Man:
Echo sounders, hovercraft and nylon nets.
A ghostly deadness writhes among the whitened dunes
A whisper on the Harmattan -
   'They've changed her name you know.
   They call this waste *The Chad Depression* now'.


*Sylvia Sikes*
*1970*

# FOREWORD

## (WRITTEN FOR 'LAKE CHAD', Published 1972)

## by SIR VIVIAN FUCHS

Chad is a remarkable lake, both in its ancient origin and its fluctuating behaviour through the ages; but more than this its influence on the local environment and the people has been, and still is, of paramount importance.

Four times, between 1955 and 1970, Dr Sikes has visited it, and here she presents an attractive combination of adventure and descriptive science which is indeed comprehensive.

The shallowness of the lake, for its average depth is only between one and two metres, together with floating islands and old submerged sand dunes, made any voyage a precarious undertaking. The accounts of her various adventures give an excellent feel of the strange environment, which cannot be matched anywhere else in the world.

For those interested in natural history there is a detailed account of the origin and history of Chad and there are separate chapters about the plants and animals. The last four deal with the people themselves, their history, their present mode of life and a look to the future. It is fascinating to read of the impact which modern technology and behaviour are having on the traditional scene, how the needs and attitudes of the people are adjusting to new methods, yet how much of the past survives.

Here too, in this small part of a great continent, we find Man short-sightedly destroying so much of the environment in the short term interest of his day-to-day survival. No longer are the fish in Chad as large as they were, but nets of smaller mesh still catch as much total weight as before. This can only lead to the destruction of the resource. The animals are killed, the vegetation is destroyed, and the development of roads is leading to greater and greater demands because of the wider distribution of products.

Only through education and inter-government co-ordination can order be brought to this cumulative chaos. The past ways of life are being overtaken by rising population and the rapid communications that modern life produces. We must hope that the Chad Basin Commission has been established in time to guide development in the right directions.

Lake Chad has been there for thousands of years but its existence, and that of all the life which depends upon it, is finely dependent upon climatic fluctuation, and no one can forecast how rapidly this may occur. Perhaps even now

*Introduction*

there are signs that the area is again on the way to the greater aridity which clearly existed from time to time in the past.

Dr Sikes presents the evidence, describes the present scene and turns our thoughts to what the future holds for Lake Chad and all who depend upon its future. Man must indeed use his wisdom and foresight if the balance of Nature is to be safeguarded in the Chad Basin. We must hope that the prologue, which looks a thousand years ahead, may not prove true.

*1970*

# Contents

| | |
|---|---|
| *Prologue* | *iii* |
| *Foreword by Sir Vivian Fuchs.* | *iv* |
| *Author's Preface and acknowledgements.* | *xviii* |

| | |
|---|---|
| **PART I: THE LAKE** | **1** |
| Chapter 1. A desert of water | 3 |
| 2. A yacht on Lake Chad | 38 |
| 3 Lagoons, dunes and floating islands | 59 |
| 4. The puzzle of the puddle. | 87 |
| **PART II: THE PLANTS AND ANIMALS.** | **135** |
| Chapter 5. The flora. | 136 |
| 6. The fauna. | 155 |
| **PART III: THE PEOPLE.** | **214** |
| Chapter 7. Conquest, bloodshed and tyranny. | 215 |
| 8. The Yedina: pirates of the papyrus. | 237 |
| 9. Fish, livestock and markets. | 265 |
| 10. Modernisation and enlightenment. | 299 |
| **PART IV: 30 YEARS ON 1973 to 2002.** | **317** |
| 11. *Appendix 1.* The fragile lake (1973/4) | 318 |
| 12. *Appendix 2.* 1975 to 2002. | 339 |
| *References* | 359 |

*Introduction*

## ILLUSTRATIONS: Photographs.

*All illustrations are by the Author, unless otherwise acknowledged.*

**PLATE 1: My 1955 visit to Lake Chad**
1. Papyrus canoes.
2. On the lake in a papyrus canoe.
3. In Dikwa town.
4. Poling through flooded fields.

**PLATE 2: The yacht *Jolly Hippo* on Lake Chad.**
1. Arrival at the lake.
2. *Sportyak* dinghy on rough waters.
3. *Jolly Hippo* with *kadais* and Yedina fishermen.
4. Mohammedu.
5. Binta.
6. Expedition base.
7. *Sportyak* dinghy with passengers.

**PLATE 3: Lagoons, dunes and floating islands (A).**
1. Lake Chad:a 'desert of water' 1970
2. Sky and water merge.
3. *Phragmites* grass.
4. Floating *Phragmites* island.
5. Silhouette of trees - navigation recognition points.

**PLATE 4: Lagoons, dunes and floating islands (B).**
1. Between floating islands.
2. Inside a floating island.
3. Broken papyrus stem.
4. Papyrus head.

**PLATE 5: Lagoons, dunes and floating islands (C).**
1. Yedina fisherman in *kadai*.
2. The author discussing a map with Yedina fisherman.
3. Dawn on the lake.
4. Sky and water merge.
5. Sunset on the lake.

**PLATE 6: Lagoons, dunes and floating islands (D).**
1. The mighty storm cloud.
2. Hurrying ashore: storm threatens.
3. Crossing a shallow lagoon in calm weather.

**PLATE 7: Flora and Fauna (A).**
1. Papyrus and convolvulus.
2. - do -
3. Ambach and water lilies.
4. Doum palm.
5. White herons, convolvulus and ambach.
6. Pelicans in flight.

**PLATE 8: Flora and fauna (B).**
1. Ground hornbill.
2. Otter.
3. Pet stork.
4. Pet ostrich.
5. Abdim's stork.
6. Sitatunga bull.
7. Female sitatunga calf.

**PLATE 9: Flora and Fauna (C).**
1. Red-fronted gazelle.
2. Wild oribi.
3. Baby monkey and leopard cub.

4. Leopard cub and red-fronted gazelle calf.

5. Leopard cub eating pigeon.

6. Leopard playing.

**PLATE 10: Flora and Fauna (D).**

1. and 2. Leopard and monkey relationship.

3. Tiger fish.

4. Nile perch.

5. Puffer fish deflated.

6. Puffer fish inflated.

**PLATE 11: Fishing (A).**

1. Bringing in nets.

2. Fish net repairs.

3. The fishing fleet at Portofino.

4. Preparing fish for drying.

5. Fish on drying platforms.

**PLATE 12: Fishing (B).**

1. Small whole fish drying.

2. Fisherman on ambach plank raft with gourds.

3. Heavily laden freight canoe with sail.

4. Fishermen at dawn.

6. Fisherman on gourd boat.

**PLATE 13: Yedina life (A).**

1 - 5. Yedina with their *kadais.*

**PLATE 14: Yedina life (B).**

1 - 6. Building a *kadai.*

7. A *kadai* after 2 years' use.

**PLATE 15: Yedina life (C).**

1. Fishing camp on floating island.

2. Noah welcomes us to his floating fishing camp.

3. Women and children in floating island camp.

4. Boys have meal in floating island camp.

**PLATE 16: Yedina life (D).**

1. Milking cow.

2. *Kourri* calves on dune island.

3. *Kourri* bull.

4. *Kourri* cow and calf.

**PLATE 17: Yedina life (E).**

1. Mixed breed cattle swimming in 1974.

2. Dune island scene.

3. Yedina hut.

4. Transporting Yedina hut in 1974.

5. Kanuri hut.

**PLATE 18:. People on and around Lake Chad (A).**

1. Musician at Kukawa.

2. Boy musician on dune island.

3. Mounted piper at ceremonial.

4. Horn blower.

5. Mounted drummer.

6. Alhaji Abba Sadiq at Kukawa.

7.  - do -

**PLATE 19: People on and around Lake Chad (B).**

1. Yedina child.

2. Island girl.

3. Schoolboy.

4. Fulani lady.

5. Kanuri ladies.

6. Hausa lady drawing water.

7. Shuwa Arab lady riding an ox.

**PLATE 20: People on and around Lake Chad (C).**

1. and 2. Yedina ladies.

3. Tuareg cameleer.

4. Fisherman.

5. Trader.

6. Fulani camel trader.

7. Farmer.

8. Fisherman.

9. Young Fulani cameleer.

**PLATE 21: Transport, Livestock and Markets (A).**

1. Lorry *(mammy wagon)*.

2. and 4. White camel.

3. Camel train - thin camels.

5. Horseman.

6. Bornu ox beside *Calotropis* (sodom apple) plant.

**PLATE 22: Transport, livestock and markets (B).**

1. Hobbled horse and view of Maiduguri Market 1955.

2. Makeshift market in 1974 on lake shore near Baga.

3. Grass mats for sale.

4. Natron blocks for sale.

5. Ceremonial mounted "jester".

6. and 7. Horses dressed in ceremonial robes.

**PLATE 23: Transport, livestock, and markets (C).**

1. Artesian bore hole.

2. Animals in shade.

3. Livestock at water hole.

4. Cows: dead and dying.

5. and 6. *Shaduf* irrigation.

**PLATE 24: Drought 1973/4.**

1. 1973 map of Lake Chad.

2. Dry papyrus.

3. Intense fishing.

4. *Merry Mermaid* and dinghy.

5. Imported heavy plank canoes.

**PLATE 25: 1974.**

1. and 2. Hand cut channels through lake vegetation.

3. Hard work poling through vegetation.

4. Square-rigged sails on canoes.

5. Yedina camp in drought on dune island.

6. and 7. Fish dam and trap on Yobe River (1970).

**PLATE 26: Satellite views of shrinkage of Lake Chad.**

1 to 6: 1963, 1973, 1987. 1997, 2001, 2002.

**PLATE 27: January 2002 (A).**

1. The Great Barrier (satellite view)

2. Hadjir el Hamis rocks.

**PLATE 28: January 2002 (B).**

1. Scattered dune islands around Bol.

2. Dyke separating polder from lake.

3. Bol peninsular.

4. Bol town.

5. Water tank and solar power panels.

6. Donkeys on 'road' (N'Djamena to Bol).

## PLATE 29: January 2002 (C).
1. *New Lycée* at *Bol*.
2. Old secondary school at Bol.
3. Primary school and children.
4. Village church.
5. and 6. Women and men after Sunday church service.
7. Heavy powered freight canoes.

## PLATE 30: January 2002 (D).
1. and 2. The *imam* and the mosque on island near Bol.
3. *Kourri* cattle.
4. The Butcher in Bol.
5. The waterfront in Bol.

## PLATE 31: January 2002 (E).
1. Oil pipeline.
2. Lake level gauge at Bol.
3. Irrigation channels at Bol.
4. An imitation *kadai* made of corn stalks.
5. A fisherman sets his net.
6. The local ferry.
7. A very small fish trap near Bol.

## PLATE 32: Missionary work.
1. Mission Aviation Fellowship float plane.
2. Model plane made by island children.
3. New mission station to be built on Haikalu island. 1969.
4. Pastor, church and members on Kinjeria Island, 1974.
5 Need for medical clinics: a man with leprosy.

**PLATE 33: Town contrasts.**

1. Baga town 1969/70.
2. Fort Lamy (N'Djamena): the great Chari River in 2002.
3. In Fort Lamy 1955: ivory seller.
4. A building in Fort Lamy, 1957.
5. Inside a church in N'Djamena, 2002.

**PLATE 34: Here and there.**

1. The Chari river.
2. Cast-net fishing on the Chari.
3. Plank canoe on the Chari.
4. Fisheries' research vessel.
5. The forge at Bol.
6. Cracked and contaminated soil.

**PLATE 35: Mainland Savannah.**

1. and 2. Crowned cranes.
3. Fish eagle.
4. Ratel (honey badger).
5. Village chief on Bornu horse.
6. Bahr-el-Ghazal horse.

**PLATE 36:**

1. Line of baobab trees.
2. Huge bull elephant.
3. Playful young bull elephant.
4. West African giraffe.
5. Lesser bustard.
6. Vultures.

Introduction

## MAPS.    Page no.

1. *END PAPERS:* Lake Chad and its surroundings.
2. Surface relief of Africa, showing Chad Basin.    xxiv
3. Chad basin, Mega-Chad, and modern Chad.    4
4. Route across Nigeria taken by 1969/70 expedition.    43
5. Baga and Portofino harbours.    83/84
6. Topography of Lake Chad.    94
7. Geomorphological units around Lake Chad.    113
8. Main dune fields.    116
9. Great, Medium and Little Chad.    121
10. Inundation zones and water table surface.    125
11. Salt deposits.    131
12. Suggested area for a wildlife park.    211
13. Kanem and Bornu under the Sefuwa dynasties.    218
14. 1974: route of the *Merry Mermaid* around the lake.    325
15. Water replenishment scheme for Lake Chad.    355

## DIAGRAMS.

1. The *Jolly Hippo* yacht on Lake Chad.    79
2. Mean surface temperatures.    91
3. Mean evaporation.    91
4. Sample echo sounder traces of the *Jolly Hippo*.    96-98
5. Hippo glares at us.    100
6. Surface level oscillations of Lake Chad.    119

| | |
|---|---|
| 7. Aquifers. | 122 |
| 8. Wind regimes. | 127 |
| 9. Baobab trees. | 141 |
| 10. Tilapia. | 156 |
| 11. Python. | 164 |
| 12. Trunkfish. | 165 |
| 13. Elephants. | 167 |
| 14. Sudan bustard. | 172 |
| 15. Terns. | 176 |
| 16. Dama gazelle. | 180 |
| 17. Scimitar oryx. | 180 |
| 18. Addax. | 181 |
| 19. Sitatungas in swamp. | 185 |
| 20. Growth curves of leopard *Pumpkin*. | 192 |
| 21. Gillnet. | 267 |
| 22. Average oscillations of the lake surface 1870 -2002. | 340 |
| 23. Maize and wheat from Bol polders. | 345 |

*Introduction*

# AUTHOR'S PREFACE

This book is about Lake Chad, an extraordinary, once vast freshwater inland sea, situated in the southern edge of the Sahara desert. This edition is a revision of my original book 'Lake Chad', published in 1972. In that book I recounted my experiences of the lake from my first visit in 1955 to the end of my 1969/70 expedition, trying to convey something of the 'feel', the 'atmosphere', of this peculiar, beautiful and challenging lake. The 1969/70 expedition resulted from my winning the *Guardian/Eyre* & *Spottiswoode* competition for would-be explorers with my plan to explore Lake Chad by yacht.

In this edition, Chapters 1 - 10 are a revised account of that expedition. Since then much has changed, including place names and situations, and the very topography of the lake itself. Over the past thirty years the lake has shrunk to a fraction of its 1969/70 size (then about 22,000 square kilometres in area). In Appendices I and II, I have tried to describe those changes as I personally experienced them in the years 1974 and 1976, and how they can be seen today in 2002 in NASA's 'Landsat' images.

In January 2002, Miss Jane Sutton who had visited the lake previously with the Mission Aviation Fellowship and again with me in 1976, agreed to revisit the lake on my behalf: Some of her photographs are shown in Plates 27 - 31. I am most grateful to her for this contribution.

I first visited Lake Chad in 1955 - a visit charged with hazard and adventure. Again in 1957, and then in 1962, curiosity drew me back. On all of these occasions I was accompanied by my friend and field assistant, Malam (Mr) Mohammedu Shehu, and, on the first and the third times, by Miss Edith Jackson and, also on the last by Miss Erika von Bernuth, to whose memory this book is dedicated. On each of these visits, we were limited to the use of canoes operating in lagoons within about 8km of the shore, and it was frustrating not to have a suitable vessel such as a yacht in which to explore the open reaches to the islands and farther shores. I wanted to take up the challenge of navigating Lake Chad's un-chartable waters, and of meeting the 'pirates of the papyrus' on friendly terms in their own homes and floating camps. So for six years I cherished this seemingly far-fetched ambition with little hope of ever achieving it.

It was in 1968, however, in the depth of the Canadian winter, that I noticed an unusual advertisement in my airmail *Guardian* newspaper, inviting applications for an Expedition Bursary from would-be explorers. Applications were requested, complete with a 5,000-word programme, for the *Guardian/ Eyre & Spottiswoode Bursary,* of which the purpose (so the announcement stated) 'is to encourage the spirit of exploration and scientific investigation.' It continued: 'The Expedition Bursary will be granted to the applicant submitting an idea for the most interesting expedition (in the opinion of the judges) concerned with exploration, archaeology, sociology, anthropology, or any other endeavour approved by the judges.'

I was the applicant privileged to receive the award for my proposed Lake Chad Expedition. It is my pleasure to acknowledge here my appreciation to the sponsors and judges of the Competition for the Award, thus providing me with the impetus and part of the means for fulfilling my ambition to take a yacht to Lake Chad. At the time of making application for the Expedition Bursary, my hopes were enthusiastically shared by Miss Erika von Bernuth, and indeed she would have accompanied me on the Expedition, taking full responsibility for all the logistics and public relations work. Two days after receiving the news of the Award, however, Erika von Bernuth was killed in a riding accident in the Canadian Rockies. She was an outstandingly happy, adventurous, determined and able person, and her place on the Expedition could not be adequately filled by anyone else. In the event, I personally undertook the work which she would have done, with an inevitable curtailment of some of the other activities planned. Mohammedu Shehu was, happily, free to join me from early July 1969 to March 1970 and thus worked as Senior Assistant throughout the period spent actually at Lake Chad. In his loyalty and reliability he was a tower of strength to me, and his friendship unfailing. One of his wives, Binta, with their elder daughter, Hanné, also joined us on the Expedition, and they took charge of the base camp at Minetti near the lake shore. I am deeply grateful to them for their help and companionship.

*In 1970, I wrote:*

There are many odd things about Lake Chad. It really consists of nothing more romantic or photogenic than a vast shallow puddle, lying in a wide depression in an even wider desert. But even a puddle may attract one's curiosity, especially when it is as unusual in character as this one *(Map 2)*.

Historically, Lake Chad is the remnant of two successive much greater inland seas : first Palaeo-Chad and then Mega-Chad. The known history of its people goes back to about the eighth century A.D., and it is a tale of conquest, bloodshed, tyranny and intrigue. The Lake is the home of a unique tribe who call themselves the 'Yedina', but who are nicknamed the 'Buduma' (people of the grass) by the people of modern Bornu, and who were formerly referred to generally as 'the pirates of the papyrus'.

*Introduction*

Lake Chad is a hydrological puzzle. The general level of its surface rises and falls by irregular amounts with a variable periodicity. These changes apparently depend upon the combined effect of three separate cycles: first, major cycles of which each comprises an indeterminate number of years; secondly, minor or seasonal cycles; and thirdly, temporary, wind-induced and tidal cycles.

Although its surface area (due to these changes in surface level) has varied between 13,000 and 22,000 sq. kms.[1] over the past 150 years, and its average depth between 1 and 2 metres; although it is recharged by only four rivers; although it has no outlet; and although it is constantly subjected to the effects of both the scorching heat of the tropical sun and the desiccating desert winds, the waters of Lake Chad remain sweet, and the lake does not completely dry up. *(But now see chapter 12).*

But perhaps its greatest fascination concerns its lagoons and islands, and their inhabitants. The shores of Lake Chad are generally fringed with lagoons of floating vegetation and studded with innumerable islands. Many of these islands consist of floating rafts of vegetation and may be either anchored by the roots, or drifting. Other islands consist of fossilised, submerged dunes or bench islands; and many consist of dune islands which are real, genuine desert islands, complete with palm trees.

---

[1] The metric system is used throughout this book

The islands cannot be counted nor can the lake be properly mapped or charted, due to its constantly changing level and the unstable character of its floor and its islands. The 'floaters', in any case, are constantly reproducing themselves by vegetative growth and subsequent fragmentation. Some people say there are currently between one thousand and two thousand islands, counting the submerged ones, but, at best, any figure must be based on mere guesswork providing only a temporary estimate.'

From late June to the end of November 1969, Mr Richard H. Cansdale assisted with the tedious work of towing the yacht the 2,080km (1,300 miles) from Lagos to the lake shore, as well as with preparations at the base camp and the launching and early trial runs of the yacht, the *Jolly Hippo*. He also assisted with the collection of equipment for the Expedition while still in England, and took two of the photographs reproduced in this book *(Plates 15.3, 23.4)*. His assistance in these ways is gratefully acknowledged. Lieutenant-Colonel and Mrs H. E. Barlow, of Jos, Nigeria, extended the warmest and most generous hospitality to all members of the Expedition at intervals during our stay in Nigeria and, in particular, to me during my convalescence from a severe illness contracted at the close of the Expedition at Lake Chad.

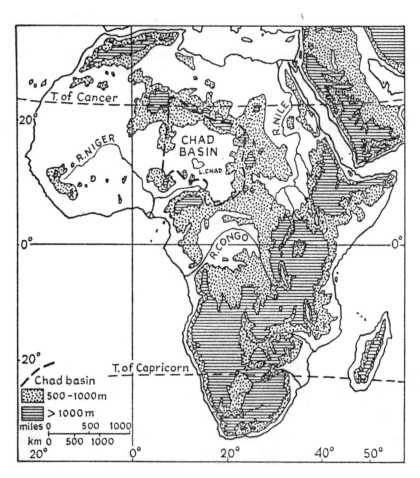

Map 2. The surface relief of Africa, with the position of the Palaeo-Chad Basin indicated.

It is quite impossible to express my thanks adequately to them. It was particularly delightful that both they and their relative, Mr Frank Drew, of New Zealand, were able to make brief visits to us at the lake, put out 'to sea' with us in the *Jolly Hippo,* and even permitted me to lead them, on foot, to within fifty yards of a herd of eight elephants, which were slopping and slurping in a muddy lagoon. *Plates 10.1 and 10.2* were taken by Mrs Barlow, who has kindly permitted their reproduction here.

Mr Drew Barlow, their son, then on vacation from college in England, also assisted in many practical ways during the fortnight in which the Jolly Hippo was launched and subsequently washed aground on a mud bank in a storm. I am most grateful for the time and effort he gave to help us.

The constant encouragement and assistance given by Miss Mary E. A. Thomas, of the University of Southampton, throughout the entire period of the Expedition, including its early preparation as well as its aftermath, was an essential ingredient in its achievement. It was thus a particular pleasure to have her fly out to Nigeria for a month's visit over Christmas 1969, in which she bravely sailed Chadian waters with us She even managed to include in her itinerary the experience of being unintentionally stranded on a genuine desert island with Mohammedu and me, and of being subsequently rescued by amphibian plane and dugout canoe. She drew the beautiful 'diagrams' of the *Jolly Hippo (1),* the glaring hippopotamus *(5),* baobab trees *(9), python (11),* elephants *(13),* terns *(15),* and the sitatungas in the swamp *(19)* and also the painting *(Plate 6.1).* I am so grateful to her for this lovely contribution to the book.

*Introduction*

Our combined gratitude goes to the pilot of this plane, Mr Ernie Addicott, and to his three passengers who put a great deal of effort in rescuing us and in attempting to repair the yacht. My appreciation also to Mr Walter Utermann and Mr Krebs of the Mission Evangelique, Haikalu, who later not only rescued the *Jolly Hippo* but also repaired it.

I also gratefully acknowledge the kind assistance and hospitality of the Mission Aviation Fellowship, and TEAM missionaries in N'Djamena and Bol in Tchad, who helped so much with Jane Sutton's visit on my behalf in January 2002. Thank you all so much. Many local officials and lake residents, Nigerian, Camerounian and Tchadian also helped me in innumerable ways: my thanks to them all.

Thank you also to Mr A.J.Hopson, Revd. Bro. Barrington, and Mr Alan Chilvers and colleague for information recorded in the Bibliography as 'personal communication'; also to Mrs Monica Haberkamp for Plate 30.3. Thanks also to the staff of the Lake Chad Basin Commission (LCBC) for help both with the original and this revised edition of my book.

The 'Landsat' images (plates 26 and 27) are published here courtesy of the NASA Goddard Space Flight Centre: I appreciate their kind permission. Thank you also to Jane Sutton who helped to shape up the revised manuscript, and to Mrs Cynthia Pocock who has so kindly proof-read it for me.

# PART 1

# THE LAKE

*Chapter 1 - A desert of water*

# Chapter 1.

# A desert of water

If I had lived about 55,000 years ago and been able to hijack a satellite, compelling it to orbit the Earth around latitude I4°N, I would have noticed a huge inland sea occupying the south-western portion of what we now call the Sahara Desert. The surface level of this inland sea was about 380-400m above sea level: it was *Palaeo-Chad* .*

Another similar trip in orbit (during the era which followed a series of subsequent arid periods when the waters of the inland sea receded and later rose again in the glacial period following), say about 22,000 to 12,000 years ago, would have revealed a new inland sea, or swamp, with its surface now just under 320m above sea level. This was Mega-Chad. It was about the size of today's Caspian Sea and occupied the central depression of a catchment area extending from the present Ahaggar to the north-west, Tibesti to the north, Ennedi and Jebel Marral to the east, the Massif des Bongos and the Jos Plateau to the south, and Aïr to the west. Another lake and swamp occupied today's hopelessly arid desert of Ténéré,

---

* Tchad (french); also the name of the Tchad Republic; Chad (english).

## Chapter 1 - A desert of water

while another huge expanse of water lay west of Timbuktu, referred to nowadays as Lac d'Arouane *(Map 3)*. Mega-Chad changed relatively little in extent until about 7,000 years ago, when it began to shrink again, parts of its former bed now being inhabited by Neolithic Man, while the Palaearctic flora to the north was superseded by the Ethiopian flora moving in from the east. The development and persistence of Mega-Chad must have required sixteen

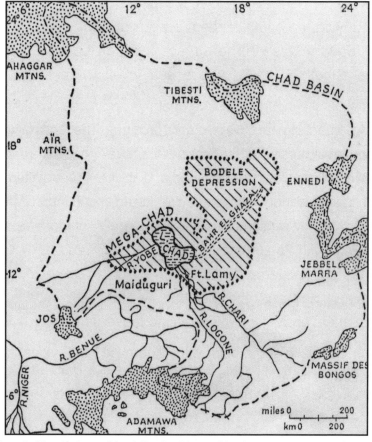

Map 3  The Palaeo-Chad Basin, Mega-Chad and 1970 Lake Chad

times the present intake of water into the catchment basin, the greater inflow probably being then, as now, mainly from the south. The flora of the surrounding country was mainly Mediterranean, although possibly there was a transitional swamp area between this and the open water consisting of areas subjected to seasonal and other cyclic inundations.

The continued existence of the Chad Basin for so long - i.e. since Tertiary (Pre-Recent) times – without any disturbance to its geological structure perpetuated its topographical characteristics to such an extent that the former limits of Mega-Chad are readily discernible even today as a strandline taking the form either of a single ridge or, in places, of a ridge complex, varying from 320-330m above mean sea level. This may be traced over a distance of 1,200kms. The fact that the same level obtains at the 'zone of capture', or relatively low watershed, between the Logone and Benue River systems, also indicates that Mega-Chad overflowed and inundated in a predominantly westerly direction.

During the latest major shrinkage period of Mega-Chad, lesser lakes persisted in some places in pans and depressions fed by rain-water and ground-water seepage, thus continuing to support much of the vegetation existing then. So, up to some point within the past 5,000 years, the western Sahara, which is now hardly capable of supporting

## Chapter 1 - A desert of water

any vegetation at all must have been in sufficiently good condition for scrub and dry woodland to flourish right across to its southern edge, although possibly somewhat discontinuously, and probably dependent upon major drainage lines.

Artefacts indicate that Neolithic Man and his livestock moved widely around the Sahara at this time without encountering any major barriers. But today Lake Chad is small compared with its great predecessors. Nowadays it apparently never exceeds 22,000 sq.kms. in surface area, and within the past 150 years has been known to decrease to as little as 13,000 sq.kms.*

In the first decade of the present century, Dr Falconer[1] crossed Nigeria on horseback, his ultimate goal and ambition being to reach the remote, famous Lake Chad. He described his moment of achievement thus:

'So this was (Lake) Chad, the object of my long anticipation! And all there was to show for it was a little brown and stagnant water amongst the reeds and rushes. The open lake (if lake there was) was securely hidden from my view. I turned away with a sense of disappointment, and thought of how Barth,[2] in very similar circumstances had 'strained his eyes in vain to discover the glimmering of an open water in the distance' and at length retraced his steps, consoling

---

* But see Ch.12

himself with the thought that 'he had at least seen some slight indication of the watery element'.

Some sixty-odd years earlier, Major Denham,[3] Dr Walter Oudney, and Lieutenant Hugh Clapperton had approached Lake Chad from the north-west, and on 4 February 1823, reached the town of Lari on the then northern boundary of the great Kingdom of Bornu. Their first view of the lake was described with more enthusiasm than either that of Barth or that of Falconer.

'Beyond… was an object full of interest to us, and the sight of it produced a sensation so gratifying and inspiring that it would be difficult for language to convey an idea of its force and pleasure. The great Lake Tchad, glowing in the golden rays of the sun in its strength, appeared to be within a mile of the spot on which we stood'.

Moving southwards, they noted that the soil changed from sand to clay, clothed with grass and dotted with acacias and other trees of various species, amongst which grazed herds of antelopes, while guinea-fowl and 'doves of Barbary' flew overhead.

'The waters rise to a considerable height in winter' (i.e. the dry season), continued Denham, 'and sink in proportion in summer. The lake is of fresh water, rich in fish, and frequented with hippopotami and aquatic birds. Near its centre and to the south-east are islands inhabited by the 'Biddomahs' (Buduma), a race who live by pillaging the people of the mainland.'

In his 1904-5 expedition, Lieutenant Boyd Alexander[4] brought a steel boat suitable for poling through the shallow Chadian waters. It was brought to the lake shore by way of Kano and the intermittently flowing Hadejia-Yobe River. Launching it on the lake, they attempted to pole south-eastwards, but failed to find a passage through the floating vegetation known nowadays as the 'Great Barrier'. They were forced, therefore, to make a north-easterly circuit instead, eventually turning west again and finally arriving, somewhat fortuitously it seems, at their starting point, about a week later.

Alexander's description of his view of Lake Chad, as experienced during this trip, reflects its remoter moods, its power and vastness, and its mystery.

'To view the real Lake Chad, Fancy must go clad in sober grey, with earth upon her head, and she must not fear to go alone, for there is a desert of water, as well as of sand. If Chad is never a gay pageant of blue and green and gold, it is often a tender vision of grey and silver, the harmony in which the spirit of loneliness abides. For loneliness is the spirit which haunts the lake if for long one follows her ways . . . There is a very mystery brooding over the place, that seems to have thrown an enchantment over its waters and peopled its islands with spirits from the shadow world.'

Poling across the open water, and past reed islands, Alexander described vividly his first day on the lake:

'Throughout the day the sun had been intensely hot, and beat upon us with its doubled power, as its rays were thrown up with a fierce glare from the water. The mirage hung upon the lake, blurring the outlines and magnifying the masses of the headlands beyond. As the sun sank past the meridian, the mirage vanished and the island headlands that had appeared shadowy and phantasmal in its thrall now stood out sharp and clear against the opposing light. But their show of substance merely mocked us, for one after another, as we touched, proved to be merely a mass of reeds standing in the water, with no rest for the sole of one's foot.'

Towards the close of this exhausting circuit of the northern part of the lake in their steel punt, Alexander records :

'We had now spent a week upon the lake, working hard to make progress, take observations and mapping the route the whole way. Yet we were completely in the dark as to our whereabouts, and not much nearer than when we started to a better understanding of the lake.'

Since Alexander's expedition to the lake, many other Europeans have visited it for a variety of reasons. Most did so in the course of their work as administrators, surveyors, soldiers, police officers, missionaries, agriculturalists and veterinarians, and a few educationists. Some saw in Lake Chad a commercial opportunity, others an opening for scientific or ethnic research, and yet others a field for missionary service to provide for the spiritual and medical needs of the lake people.

The lake and its environs claimed a high proportion of lives among these adventurers, for many of whom neither the cause of death nor the place of their graves is known. Nevertheless, all developed their own views of Lake Chad as they grappled with maps that were, and still are, incapable of being accurate, carried out generally self-appointed tasks, speculated recklessly as to what was real and what was merely mirage, and struggled to survive in conditions that conspired quite positively against survival.

My own visit to Lake Chad in December 1955 was brief, but nonetheless provided me with a remarkably comprehensive experience of the lake. On that occasion I was accompanied by Miss Edith Jackson, a lecturer at the Nigerian College of Arts, Science and Technology at Zaria, in Northern Nigeria, and Mohammedu Shehu, hunter-tracker and field assistant. We travelled by road from Zaria to Maiduguri, the administrative capital of Bornu. Bornu is the area that embraces the southern part of Lake Chad, while Kanem is the area that embraces the northern part.

Bornu today (1969/1970) ranks as a province of North East State, one of the twelve federated states which comprise the Republic of Nigeria*. The British reinstated the traditional ruler of Bornu, called the Shehu (or Sheik) soon after the defeat and death of the infamous Sudanese brigand Rabeh in 1900. Rabeh had advanced from the north-east with his private army and conquered the already powerful and extensive Bornu Empire in the latter part of the nineteenth century. Then the French, British and Germans set up colonies, such that the French took over Kanem, the Germans a minor part of Bornu, and the British the remainder of Bornu. The German part, Cameroun, was subsequently divided between the French and British. The Shehu today is still the head of all local government in Bornu, administering both his capital city Yerwa (a fortified city inside the present larger town of Maiduguri) and outlying districts, through a structure called the 'Native Administration'.

Bornu is traditionally the home of the Kanuri[†] tribe. They are people of mixed stock, basically comprising descendants of the Sefuwa-Kanembu, who in turn are of Berber and Yemenite Arab stock.[‡] When the Sefuwa-Kanembu migrated southwards between the fourteenth

---

* Bornu today (2002) is a Nigerian State, one of many.
† Pronounced 'Kah-noory'
‡ The history of Kanem-Bornu is more fully described in Chapter 7.

and nineteenth centuries they absorbed, through intermarriage and slavery, most of the indigenous tribes they encountered. Their language is thus as diversified as their physique and origins. Kanuris often say that one clan may find it difficult to converse with another distantly related clan, as the dialects differ so much.

Wherever one meets Kanuri, however, one has the impression of self-confidence, a direct manner, and a pleasant way of looking one straight in the eye. Their life is always closely bound up with the importance of their transport animals: horses and donkeys, horses holding a place of special honour. Camels are also used by the Kanuris of the northern parts of Chad, but are generally handled and maintained for them by the Fulani, a pastoral tribe who traditionally migrated southwards deep into Nigeria in search of dry-season pastures each year.

It was the Fulani who conquered the northern Hausa Kingdoms under Usman (Othman) dan Fodio in his *jihad* (holy war) in the early part of the nineteenth century, and it was the Fulani who drove the Sefuwa-Kanembu south in 1812. The Fulani were thus the traditional enemies of the Kanuri. Under Rabeh, however, they too had to pay tribute, and today they enjoy a fairly harmonious existence alongside the Kanuri in Bornu.

Shuwa Arabs also form a sizeable population in both modern Kanem and Bornu. They too are traditionally nomadic, dependent upon their livestock. They are experts with their horses, and their main pack animal is the ox. Hausa settlers from Sokoto in the north-western part of Nigeria may nowadays also be found near the lake shores, and Hausa traders in most of the markets. The settlers at the lake shore are mainly fishermen and businessmen. The fishermen of the actual lake are mainly the Yedina tribe,* but nowadays certain Kanuri clans have adopted a way of life very similar to the Yedina, and even some of the Sokoto Hausa fishermen who venture out to the nearer islands.

They would certainly not have been able to do so prior to the advent of the French, who introduced two steamers as patrol boats on the lake, at the beginning of this century, to control Yedina piracy. Today (1969/70), however, the inhabitants of the lake itself and its islands seem to be among the most peaceful and hospitable people one could wish to meet anywhere.

In Maiduguri we saw people belonging to all these tribes, as well as a considerable population of Europeans, and a number of French-speaking African traders from the Sudan and the countries now known as Tchad and Niger. In the market we were struck with the tendency for

---

* Decribed in more detail in Chapter 8

merchants with specific skills and trades to congregate in particular areas: dealers in saddlery for camels and horses in one place; tailors (each with his own treadle sewing machine and a display of his work) in another, and money-changers in yet another. The latter dealt in every conceivable type of currency including both genuine silver Maria Theresa dollars* and aluminium copies. They carried out mental calculations of their transactions with lightning speed, and profiteered with aplomb. There were hatters, butchers, blacksmiths, mat-makers, greengrocers and sellers of brass and silver ware. There were horse, donkey, camel and bicycle dealers, and sellers of Koranic literature.

Most fascinating were the dealers in medicines and charms: from them one could buy concoctions to ward off diseases and enemies, or charms to induce fertility. Bloodletting, into cupped horns on the back, was in progress, the patients sitting motionless around the practitioner. Everywhere beggars, each holding out a grubby half-calabash, sought alms: they were blind, crippled or leprous. Everywhere was dusty, flies crowding on sprawling heaps of refuse, the open sewers stinking powerfully, while men relieved themselves openly, and without inhibition, into them.

Outside of the market proper lay a lorry park, a camel

---

* Austrian coins minted in pure silver

park and a donkey and horse park. Beyond these again was the 'Government Reservation Area', formerly an exclusive European residential area, but now also including Nigerian nationals, flanked by the colonial administrative buildings and 'canteens' or shops. The Shehu's palace and administrative buildings were situated inside the city gates, within the old city of Yerwa.

The market has changed little since 1955, although the wares traded now include a high proportion of patent medicines, transistor radios and motor scooters. Outside the market, the most noticeable changes are the metalled roads, roundabouts and modern road signs, along with a noticeable dearth of European faces. Inside the administrative buildings today, an encounter with a European official is rare, almost all government posts being held by well-educated, indigenous, English-speaking nationals of Bornu and other provinces of North-East State. This situation differs radically from that obtaining in Fort Lamy, in Tchad Republic and in the Republic of Niger, where Frenchmen still hold a high percentage of the executive posts.

From Maiduguri we drove via Dikwa (Rabeh's last headquarters) to Ngala near the southern end of the lake. We had been advised to try to go from there to a village called Wulgo, said to be 'right on the lake', where we could hire papyrus canoes, see a variety of wildlife and generally enjoy an interesting holiday.

## Chapter 1 - A desert of water

As we left in the early morning for Dikwa, we passed groups of boys sitting under the shady neem trees that line many of the Maiduguri streets, in little Koranic classes, each led by a *Malam* (Koranic teacher). Each boy held a wooden board inscribed in Arabic script with a passage from the Koran which he was learning to recite. Most of the malami (teachers) were armed with a whip consisting of several cords: if a boy was inattentive and displeased his malam, he was promptly struck with the whip, which would catch him on the shoulders, chest and face. As we watched these classes, Shuwa Arab women began riding in from outlying districts to the market on their oxen, laden with wares to sell, and also carrying even lambs and kids in their arms.

Driving towards Dikwa we passed through flat country at first, with shady acacia thorn trees with reddish-yellow coloured bark. There was a good grass cover under the acacias and numerous termite hills. We saw some red patas monkeys and a grey duiker (small antelope), but no other wildlife in this thorn scrub area, apart from birds. Bee eaters and grey hornbills were abundant, and we saw two herons in the distance, but could not determine the species. In this area we passed a group of Shuwas travelling with their oxen laden with the materials necessary for constructing temporary shelters, household utensils, a few goats, sheep and dogs.

Dikwa stood out starkly on the barren sandy plain beyond the thorn scrub area. The houses were solidly built in the moorish style, with thick mud walls, domed roofs, and parapets pointed with asymmetric cones at the corners, and drained by long gutters of bent corrugated iron. The frames supporting such roofs are usually made of palm beams as these are inedible to termites *(Plate 1.3)*

The walls are usually made by means of rows, several deep, of conical 'mud pies' made of clay, laid direct on the bare soil without any kind of foundation. When the first layer has been laid (about a metre wide) clay is slapped into the interstices and allowed to set. Another layer of mud pies is then laid over the interstices in the first layer, clay slapped in between, and the line again allowed to harden. The supporting beams for the roof may be built into the walls or into mud pillars and, in very important houses, the palm beams may be pre-stressed to induce satisfying curves for the actual arch of the domed roof.

When the roof beams have been set and secured, fronds are laid across to form the ceiling, on top of which clay is again slapped and carefully smoothed over by hand. The asymmetrical corner cones usually, but not always, contain the heads of palm beam uprights. They are not entirely decorative in function, although their detail reflects the work of the individual architect-builder. In some cases they apparently also have a definite function in channelling rainwater to the gutters which run along inside the parapet, and then to those poking out through it at right angles for some eighteen inches or more beyond the mud walls. The whole building is then plastered with a special type of cement-like clay, with, in very sophisticated buildings, a mica-containing additive, giving it a beautiful shiny finish. We did not notice any of the latter at Dikwa, although houses with this type of finish are still to be found in Maiduguri.

Nearing Ngala, the road was banked up to form a causeway above the endless *firki* (black clay) flats. During my 1957 visit, cuts had been made into this causeway in readiness to breach it when the lake water and flooding rivers began their annual inundation of the plain, and in some places these made the track so narrow that we had difficulty in getting our truck between some of the cuts.

During our 1955 trip, however, there was little water in the vicinity of the road on our outward journey, and the *firki* was still baked hard, cracked into endless flakes and cores by the fierce sun and dry winds. Around Lake Chad, extending through many parts of the former Mega-Chad floor, lie these inundation areas, parts of which still become flooded when the lake makes its annual dry-season rise. The amount of this annual rise in level varies from year to year, and is related to the actual amount of rain falling within the whole area of the Chad Basin, or catchment area, during the *previous* rainy season, and particularly, in the Chari-Logone river system.

The ultimate cause, and modifying factors, determining the amount of rainfall from year to year, how much actually drains into Lake Chad itself, and what may be the basis of the cycle of variations is still unknown. When the lake rises, it overflows outwards across Mega-Chad's old floor, causing wide shallow floods. Just as with present-day Lake Chad, the floor of Mega-Chad was covered in many parts by thick black muddy 'ooze', rich in organic matter. As Mega-Chad receded, these black muddy deposits were exposed and became dried out by the sun.

However, where inundation still occurs, the mud is instantly reconstituted into a saturated, soft, fertile mass, on which crops can be grown, but through which all transport (other than by boat) is exceedingly difficult. A red sand may be found underlying the *firki* in some places - an encouraging fact to know, if travelling by car, as it means that a limit is set to the depth to which one's vehicle can sink (if one gets stuck) before obtaining the necessary assistance to extricate it.

As we crossed these *firki* plains, the sun was high and the plains shimmered in the heat so that they seemed to consist of nothing but wide dancing mirages. Far away across the plains, there was a herd of seven western, or bubal, hartebeest grazing. They seemed as tall as giraffes and as lacking in real substance as the dancing rays themselves. I suppose we and our truck looked as big as mastodons to them, for they withdrew at a mad gallop into what seemed to be the sky.

We were tired when we reached Ngala, for the road had been dry, dusty and very bumpy. We spent the evening poking about fruitlessly in the sandy ridge near the thatched house where we were to camp for prehistoric arrow heads and pottery. We admired a gigantic pot supposed to be of prehistoric So (or Sau) origin which stood on the verandah of the house, and speculated somewhat ignorantly as to its possible use. Our final conclusions as to its obvious connection with 'Ali Baba (a common name in Bornu) and the Forty Thieves' may not have been so very far-fetched, as I have since learned that similar pots were apparently used by the So for burial purposes, the corpse being stuffed inside in a folded, sitting posture.

We arranged for two horses to be brought for our use in the morning to take us to Wulgo. The people said only two would be brought although we had also asked for one for Mohammedu: we later saw the reason why they did not bother to promise one for him also. Then we pored over our maps and prepared our cameras, with one spare film each, the shotgun with a few cartridges for duck, a packet of sandwiches and a thermos of coffee. We had little realised the full implications of a remark casually dropped by a sage in Maiduguri who had said : 'The rivers are still flooding and the lake already rising, you know . . . '

After an early breakfast we were ready, waiting for our horses to arrive at 6.30a.m. There was an irritating delay before they arrived and then an excessively long wait while a discussion took place between the horse owners and the Village Head. The subject matter was not clear to us and no one interpreted. Eventually, we mounted and set off following Mohammedu and the horse owners who were on foot. We settled ourselves to the prospect of about ten or twelve miles slow ride to Wulgo.

Our thoughts, however, were somewhat rudely interrupted by the sudden disappearance of the road under water. Some canoes were tied to growing corn stalks at the roadside. We dismounted and were motioned to the nearest canoe. Meekly, but puzzled, we got in and sat down. Now we were poled across flooded fields for some distance - maybe a couple of miles - until the village of Gambaru hove in sight, standing on a low ridge just above the water level *(Plate 1.3)*. It was completely surrounded by water, and several houses at the perimeter had subsided to become mere muddy mounds in the water.

We disembarked at Gambaru and were led on to the wooden bridge that linked the former British Cameroons with the former French Cameroun, over the Bahr el Beïd (a water-course usually dry except during the rainy season and Lake Chad's inundation period).The water in the Bahr el Beïd is also said to back-flow on occasion, a common situation in the inundation area water-courses, both of the Chad Basin and the Benue River Valley. Nowadays (1970) the former British Cameroons here form part of Nigeria's North-East State, while the former French Cameroun forms the northern part of the Republic of Cameroun.

When Edith, Mohammedu and I walked on to that bridge in December 1955, however, the Bahr el Beïd was far from its usual waterless state, for the bridge was awash, while a surging muddy torrent swirled below The bridge trembled as we wandered to the far end, greeted some African Customs officials on the French side and wandered back to where some freight canoes, tied to the bridge deck, were being loaded. The surrounding countryside, as far as the eye could see, was flooded, partly submerged trees along the river banks sticking up with a somewhat questioning air about them that seemed to match our own.

Soon a large man got into one of the already well-laden canoes, pushed the loads around a bit, and then signalled to us to get in. Edith went for'ard, I amidships and Mohammedu aft. The large man sat down in the bows, paddle in hand.

Some more sacks of corn, a bicycle and the owner of the bicycle (an indigenous French-speaking Customs officer) were brought aboard, and a second paddler seated himself in the stern. We now had about two inches of freeboard, and the canoe seemed to be straining at its painter for release into that crazy surging flood rushing towards Lake Chad.

The man at the stern cast off and we hurtled away between fish traps, tussocks, tree stumps, branches and sandbanks, the paddlers rather vaguely directing the canoe to keep its bow pointing towards the lake, some miles ahead. We three were sitting facing forwards on sacks and bundles of freight, our hips firmly wedged between the gunwales of the canoe and our knees poking up in our faces.

At first it seemed terrifying, then hilarious, and then very serious. After some three hours of this agonising incarceration in the canoe and incineration by the sun, I yelled to Edith that I presumed that she also had observed that no canoes were travelling - or indeed could conceivably even try to travel - upstream. She replied rather drily that the point had not escaped her notice. There were masses of wildfowl in flight over the flooded area as we passed but, as any attempt to disinter gun or camera would have upset the canoe immediately, we could merely note and admire them in a detached kind of way.

Some time later, the water became less turbulent, the stream course better defined, and the banks no longer submerged by the flood. We coasted over towards a dried-up-looking village, a mere handful of derelict huts near two big shady trees. We were going ashore! The canoe grounded and we staggered out, flexing our aching legs to try to restore at least a little circulation. Two deckchairs with rawhide seats were placed under the trees for us, and we were able at last to get at our thermos for a sip of coffee. Some freight, including the bicycle, was now offloaded, and we re-embarked, the canoe now riding higher and more comfortably.

It was now possible to do some shooting, and we got twelve duck and two geese as we were paddled in a leisurely manner towards Wulgo, arriving there around 4p.m. The Village Head greeted us and, after some discussion (in mixed Arabic, Hausa, and pidgin English) said we should stay in the school, as the roof of the guest house had recently been blown off. The school was a large building with a corrugated iron roof, mud block walls and a concrete floor, situated on a raised ridge of sand. Two iron bedsteads without mattresses, a table, some chairs, a Persian carpet and an old, very dusty mosquito net were brought, together with an enormous wide-mouthed black cooking pot and two eighteen-litre kerosene tins containing water, which had something like seaweed or frogspawn in it.

About returning to Ngala . . . ? They just shook their heads. Could we buy some rice or guinea corn flour? They again shook their heads: there was scarcely any food left in the village. We gave away gifts of duck to the boatmen and as many local VIPs as we could identify, keeping two for ourselves and one for Mohammedu. Edith boiled our two simultaneously in the big black cooking pot, and I boiled the peculiar water in the kerosene tins for drinking, pretending it was really tea in order to sustain our morale.

We both agreed that having actually reached Lake Chad, we might as well make the most of our opportunities, so we found a local hunter to guide us to some good shooting. Although we had hoped to photograph hippos, we were led away from the lake into dry scrubland where we saw little wildlife. On our return, however, we did hear some hippos, and I shot a red-fronted gazelle. This was duly quartered and part of it distributed to our hosts, while we cooked a leg for ourselves. As darkness fell, myriads of tree frogs started up their chorus a characteristic feature of night time everywhere on Lake Chad. The mosquitoes also started up in serried ranks, wave upon wave of dive bombers, in determined accurate raids on every exposed square millimetre of our skin. Edith rolled up in the Persian carpet, and I in the mosquito net, vying with each other in rich and varied comment and unremitting combat against our aerial attackers.

## Chapter 1 - A desert of water

That night seemed very long, and we were weary when we got up early next morning, but we were quite determined to go out on the real Lake Chad. We breakfasted on duck, gazelle and 'frogspawn' water. A boatman arrived to take us to the boats. We followed him past the village, to where the papyrus and reeds began, a fringe along shore.

The path passed into the papyrus to a damp place under a tree *(Plate 1.1),* and there we saw some of the famous papyrus canoes lying in the shallows. These are called *kadai* in the Yedina language, and consist of papyrus fronds neatly cut and bound into bundles with soaked grass or palm fibre rope. Each *kadai* has a blunt stern and a gracefully curved prow, which is held up in place by a rope stay, tied back into the hull. These boats were in two sizes, some about 4m and others 5 - 8m long. A large one takes two men about two or three days to build, and must be launched immediately on completion to prevent the fronds from drying out and becoming loose. Even while being built it is kept constantly in the shallows.

Edith and Mohammedu got into one canoe and sat down, their boatman standing with a pole in the stern, while I got into another with two boatmen *(Plate 1.2)*. As we sat down on the floors of our respective canoes and the men pushed off, the canoes gradually subsided somewhat so that the water seeped in and we were each sitting in about four inches of water inside the canoes.

The boatmen punted and paddled us at remarkable speed (I later found that they can do about 3 knots with ease in one of the larger sized *kadais* if unladen and with two men punting) through papyrus-fringed lagoons. In some places the papyrus gave way to *Phragmites* reed-grass. This is a very tall reed-grass with cane-like stems and a pretty flowering head which, like papyrus, grows in the water and forms floating rafts and islands.

The papyrus of Chad, the same as that used in ancient times for paper manufacture, grows tall. It is really a rush, and its stem is triangular in cross-section. Its flowering head looks like a comical - sometimes very graceful - topknot and, when the flowers are open, has a fluffy appearance. The stems arise vertically from rhizomes, or swollen stems, which are buoyant and float on the surface of the water, and, where they are abundant, become intertwined to form a thick floating mat. Roots dangle down from the mat of rhizomes into the water and may actually anchor the mat to the lake bottom in places where it is sufficiently shallow.

In storms, or when the lake rises, some of the roots are torn out of the substrate or broken off, and portions of the mats or rafts float away as 'floating islands', or *kirta*, to fetch up later on some farther shore, or else to take root again as they pass over some submerged dune.

## Chapter 1 - A desert of water

From time to time we encountered papyrus barriers between one lagoon and another. When these were narrow our boatmen jumped on to the floating papyrus mat, helped us out, and then hauled the canoes over the mat and through the papyrus to the other side. We had to be careful how we placed our feet: a careless move and our legs slid straight through, between the rhizomes, leaving us straddling a knot of rhizomes while our legs dangled helplessly underneath. It was hot and humid inside the papyrus stands, and insects hummed around our eyes and ears. At one point a hippo grunted near my canoe, but I did not see it. There were numbers of small birds to be seen, but apparently the migrant water fowl preferred the estuary of the river and the inundation areas to the lake proper.

The lagoons here seemed to be about 3-4m deep and, when the .410 adaptor for the shotgun was accidentally dropped into the water, the boatmen dived repeatedly for it, remaining under water for remarkably long periods while they searched, unsuccessfully, for it. They were strong swimmers and apparently quite unafraid of the waters of the lagoons.

Inside the lagoons the water was calm, ruffled only slightly in the more open parts by the breeze. We never really experienced the open lake beyond the Wulgo lagoons on this occasion, although we glimpsed a very extensive stretch of water through an opening in the papyrus just before we turned back. There was something compelling about these lagoons, for here the silky, silvery-blue water, the green papyrus reflected in it, and the cloudless sky above, seemed somehow to be continuous, as if they were not separate things at all. Even these dream-craft in which we were sitting, the kadais, their design possibly dating at least from Noah's day, seemed to belong to the eternal mystique of this place, this mirage-lake.

By the time that our craft nosed back into the reedy shallows at Wulgo, the sun was high, and very hot. Our real problem now was how to return to Ngala. Not only was our vacation somewhat limited, but it was quite clear that Wulgo was very short indeed of even basic foodstuffs. We munched some more duck and gazelle, and sipped 'frogspawn' water, while the village elders conferred. About 11.00 a.m. two sorry-looking horses, ready saddled, were brought. They were heavily scarred with criss-cross brands on their faces and flanks. The saddles were in a state of abject disrepair, and appeared to remain in position by mistake rather than by intention. We mounted and set off at an incredibly slow walk, Mohammedu on foot in his chosen position as gun-bearer just in front of my horse's nose.

It was terribly hot, dazzling and dry, and the path very winding as we dawdled unwillingly between flooded areas,

## Chapter 1 - A desert of water

sticky *firki* bogs, and muddy pools. While crossing one sticky pool, Edith's horse slipped so that her precious camera and film plunged into the muddy ooze. We plodded on and on through the midday heat, until we came to the brink of a vast flooded area, with grotesque trees poking up in the middle, and a fast current rushing down through it towards the lake.

We wondered if this was the Yedseram river but were never able to confirm this. We could not see the other shore - if indeed other shore there was. Dismounting, we sat wearily on the sand with Mohammedu, sipped more 'frogspawn' water, and again munched duck and gazelle, while the horse-owners leapt into the saddles and galloped away in a cloud of dust. Naturally we wondered if we had been abandoned to our fate. The thought struck me with great force how incredibly easy it must be, around Lake Chad, to die of starvation and thirst although one may be within only a few miles of its 60 to 80 milliard $m^3$ of thinly-spread water and its rich stock of fish. For, wherever the edge of the water is, there desert immediately begins and stretches away with all its desiccating power concentrated on every living creature foolish enough to expose itself.

But we had underestimated the kindness and concern of our hosts from Wulgo village. After what seemed an interminable period of anxiety while we waited, they suddenly reappeared, dismounted, flung off both the saddles and their own clothes and led the horses straight out into the water. Soon we saw that both men and horses were swimming, swept diagonally across the current, to disappear beyond the trees sticking up through the flood water.

A little while after this minor drama, we noticed some most peculiar contraptions being pushed towards us through the shallows on our side of the flood. These were rafts of short poles, about 1.2m long, lashed together loosely, and buoyed up at the corners by either two or four huge gourds (calabashes) according to the shape of the raft. Edith was lured on to one of the four-gourd type, and told to sit down. She put her camera inside one gourd and the thermos flask in another. The gourds had their opening upwards, very conveniently, an arrangement which I later learned was connected with fishing. For normal use, the fisherman sits on the raft with his legs dangling between the poles. To propel himself, he merely waggles his legs. As he collects fish from his traps and lines, he drops them into the corner gourds. In the two-gourd type, he sits astride the poles, and balances himself as if on a sort of surf board, - a system used in other parts of the lake. *(Plate 12.2).*

An old grey-haired man and three boys pushed Edith's raft out from the shore, wading in and then swimming beside it as they pulled and pushed it across. I felt terribly anxious for her safety.

## Chapter 1 - A desert of water

Now I was coaxed on to another similar raft, just as Edith's 'ferry' returned for Mohammedu. The raft men were apparently Yedina, huge men bursting with gigantic muscles, and very strong swimmers. As they pushed my raft across, two lads, each on two-gourd craft, overtook us. Each was carrying a saddle on his head, and they were waggling their legs at great speed with marvellous balance, as they laughingly traversed the flood.

I was glad to find Edith safe and well, with the horses on the other side, and Mohammedu also soon joined us. The horses were re-saddled and we set out once more, plodding along in the blazing heat. However, our buzz of conversation and hilarity over our most recent adventure soon ebbed. We both rode asymmetrical saddles that were clearly not designed with any thought for the comfort of horse or rider. Conversation flagged and finally died. Even the ever cheerful Mohammedu seemed to be concentrating on nothing more noble than the business of putting one foot in front of the other, and keeping going.

I began to think about old-time explorers like Falconer[5] who crossed Nigeria on horseback, and concluded that he probably used a best-quality British army saddle complete with felt numna, and wore padded breeches. My thoughts ran on to those intrepid adventurers who tried to find out where the Niger began and ended, and which way it flowed: they knew all about canoes, rapids, floods, tropical sun, and suchlike, but, of course, few returned alive to describe it for others.

The sun began to sink, growing bigger and bigger as it neared the indeterminate and hazy horizon, until it looked like a huge circle of red paper stuck on the sky. Then, without warning, it suddenly became misshapen, elongated, grotesque, and then tumbled out of sight as if ashamed of its distorted appearance. In the receding twilight we saw some trees on a mound, the thatched house, and . . . the car. We were back at Ngala.

I suppose it might have been justifiable, after this experience, to feel that we had 'done' Lake Chad, but this was not the reaction of any one of the three of us. We had all tasted Chadian adventure and wanted more of it. We had come to appreciate and admire the lake people and wanted to know them better. We had discovered that Lake Chad is not just a mirage but a real, scintillating mirage-lake.

Inevitably, then, 1957 found me travelling farther around the south-eastern side of Lake Chad to the Ubangi-Chari, Logone and Fort Lamy, and 1962 found me again in company with Edith Jackson and Mohammedu Shehu, now joined by Miss Erika von Bernuth, visiting Mongonu and Ngurno on the south-western side.

## Chapter 1 - A desert of water

On this latter occasion we spent most of our time trying to see and count the famous Chad elephants which are permanently resident around this area. Indeed Mohammedu and I did have a splendid opportunity to observe the main herd of over one hundred animals for a considerable period. The lake waters were again rather high on this visit and areas near Ngurno were inundated.

The town of Ngurno is rather small and insignificant - a shadow of its former glory when the great Shehu (Sheik) Laminu made it his military headquarters in 1812 after the former capital of the Sefuwa dynasty, Kasr Kumo, was sacked by raiding Fulani warriors. Shehu Laminu repulsed the Fulani and accompanied the Sultan southwards, occupying Ngurno himself, while the Sultan settled in nearby Birnin Kabela.*

Here we also familiarised ourselves with Mongonu, and its ancient market, always held on Saturdays, rode camels and noted the sandy ridges that mark the former limits of the lake, and stand slightly above the surrounding flat ground. Again we had travelled by car, the road from Maiduguri to Mongonu and Ngurno consisting of nothing more than a double track through the sand. Friendly Kanuri horsemen clad in either white or else indigo-dyed robes, a conical broad-brimmed hat of woven grass, and a sword

---

* The detailed history of this era is decribed more fully in Chapter 7

on the thigh, trotted along beside the track, raising the right fist in greeting as we passed.

Some Shuwa Arabs on their oxen plodded along, and some exceedingly arrogant-looking young Fulani cameleers cantered past, overtaking our truck as we ploughed heavily through the deep sand. We passed women and children riding or driving donkeys, some laden with panniers containing big earthen pots of water. There was an old man on a horse driving a herd of cattle to water, and a horse with two Kanuri girls riding tandem

We passed an artesian well, one of the first to be opened in this area and saw the herds of livestock crowding around it to reach the warm water as it gushed out under pressure from an open pipe on to the sand, forming a dirty-looking pool. There were groups of cattle clustered in fragmentary patches of shade cast by thorny acacia trees, some of which had their branches slashed at the bases to bring them within reach of the scrawny cattle. Only the sheep and goats really seemed able to thrive and look fat in this arid place, with its parched khaki-coloured grass, and sandy patches.*(See Plate 23.1 and 2)*.

On this visit we made our main base at the brick Guest House at Mongonu. This was a two-roomed structure situated just outside the village among some shady trees. During the elephant hunting period however, Mohammedu and I stayed at Ngurno. There I was allocated a Kanembu-type grass 'beehive' hut *(See: Plate 17.5)*. These are made entirely of grass, with very thick walls and roof, and a door with an overhanging grass porch. Mine was floored with persian carpets, and supplied with two large pots of freshly-drawn well water. These water pots are porous so the water is always beautifully cool inside. The people are always very particular as to how water is drawn from these pots, only the special dipper being used. We never managed to reach the lake proper on this visit as the inundation areas were very extensive. ***But this tantalizing situation only served to whet my appetite yet more for the opportunity to make a prolonged visit if ever it should be possible.***

*Lake Chad versus The Sahara Desert*

# Chapter 2

# A yacht on Lake Chad

When I began my preparations for the 1969-70 Lake Chad Expedition, I gave a great deal of thought to the question of the kind of boat I needed. It should be an ocean-going vessel, small enough to trail up-country behind a Landrover over narrow roads and bridges, and yet tough enough to stand up to the squalls so well-known to Chadian residents and fishermen.

To enjoy those lovely breezes from the desert, it must, of course, be a yacht rather than a launch, yet it must also be powered to cope with the breezeless lagoons and awkward shallows. It must be capable of being handled by only one person if necessary, and yet have accommodation for up to four people. It must have a cabin with 'all mod. con.' (because it would have a mixed crew), and an adequate galley because we should be on the lake for days at a stretch. It must be steady even under sail because we would have equipment that we should not want damaged, and a draft of not more than about half a metre because so much of Lake Chad is very shallow indeed and so many of the submerged dunes lie at only about that depth below the surface.

I began to be a fanatical reader of boating magazines and catalogues, a lounger around boatyards, and a pupil at a yachting school. I even wrote to famous yachtsmen to ask their opinion and advice. I went to a place on the east coast of Britain, on a bitter day in the spring, with Richard Cansdale and his brother David, to try out a possible vessel on the high seas. The seas were indeed running high that day, but the salesman took us out in spite of the fact that the yacht was unfinished, had no safety equipment, and apparently had a lunatic at the helm. We ran more risk to life and limb on that brief voyage than during the entire Lake Chad Expedition.

And then next day I saw the craft I wanted. Disconsolate, I was wandering through the London streets trying to find various types of necessary equipment, when I saw a yacht on display in a shop window in New Fetter Lane. Going inside, I asked the salesman about it, but he evidently thought me an unpromising-looking customer, for he declined to answer my questions, and merely fobbed me off with a pamphlet.

The pamphlet, however, proved to be more informative than the salesman, and it was quite clear to me that this was the right vessel for Lake Chad. Not only so, but the price was within reach and delivery date could be guaranteed in time for shipment in June. Further enquiry revealed, moreover, that both a reinforced trailer and a protective plastic cover could be supplied in time. Then I

## Chapter 2 - A yacht on Lake Chad

went to Chichester for a run in an identical demonstration model, and again it was a chilly, choppy day. I had no doubt at all now that this was indeed the boat for Lake Chad, and I also knew instinctively as I pondered over her firm snout, her steady manner, and her broad beam that her name could be none other than the *Jolly Hippo.*

Designed by E. G. Van de Stadt, and built by Messrs Dell Quay productions of Emsworth, Hants, she was a Mirror Class, offshore yacht, 19 ft (5.8m) overall, with 7ft (2.2m) beam and 2ft (0.6m) draft, and bilge keel, powered by a Volvo Penta 7 h.p., single cylinder, marine diesel engine. She was sloop-rigged with aluminium mast and boom, and roller reefed foresail. She carried 59.5 sq. ft. (5.5 sq. m.) mainsail and 62.9 sq. ft. (5.85 sq. m.) genoa. Her cockpit had longitudinal seats capable of sleeping two crewmen, and a two-berth cabin with drop-leaf tables, flush toilet closet, and galley annexe. Her fibreglass hull was marine blue and her decks finished in white. Her registered tonnage was 2.23 *(See Plate 2.1 and 3).*

I also gave a great deal of thought to the question of dinghies, and finally settled for two identical Sportyak II, 8ft 6in (2.6m) x 3ft 9in (1.1m), built with moulded polyethylene, catamaran-style, hulls. These only weighed 85lbs (36.9Kg) each, so were light enough to transport on the Landrover roof. Each was fitted with rowlocks and oars, and could also be used under sail, or with an outboard motor. Each had an aluminium mast, boom and tipping lug with 48 sq. ft. (4.5 sq. m.) terylene sail, and the outboards I selected were

Deflectajet 3.5 h.p. motors. These remarkable dinghies are not only unsinkable but very durable, and proved equally adept at slipping through the muddy shallow lagoons as at riding the great storm waves of the open lake in the wake of the Jolly Hippo. *(Plate 2.2 and 7)*. So good were these Sportyaks that we never needed to make use of a rubber inflatable dinghy taken as a standby in case of emergencies.

On 26 June 1969, simultaneously Richard Cansdale arrived by air, and I by sea, in Lagos, Nigeria. After ten memorable days of toilsome patience, we managed to assemble all our equipment, hitch up the yacht on its trailer to the Landrover (named 'The Mammoth', also marine blue and white in colour matching the yacht) and set out on the 1,300 mile (2,080km) road journey to the shores of Lake Chad *(Map 4)*. Driving at speeds of under 40 km.p.h. we managed to keep wear and tear on the trailer to a minimum. Nonetheless some minor running repairs proved necessary along the way trailing the yacht, the *Jolly Hippo*, from Lagos to Lake Chad.

The Inspector General of Police, Nigeria, had kindly given us a police escort from Lagos. Without his help I do not think that we could have completed the journey intact, for frequently he had to halt oncoming traffic and direct it aside to enable us to negotiate particularly difficult places or narrow bridges where priority might have been in question. Later I returned part way back to collect Mohammedu from his home in Zaria, to bring him, his wife Binta, and daughter Hanné to Lake Chad.

Negotiations at Jos and Maiduguri were rather prolonged, and then there was the search for a suitable base camp, and the decision as to whether to set up a canvas camp or huts, or whether to use available buildings. The rains and gales were in full force when we arrived, and it was the wrong season for hut-making, so we looked for available buildings. It was Alhaji Abba Sadiq, District Head of Kukawa, who drew our attention to the disused Italian road-builders' camp at Mile 90 (locally pronounced as 'Minetti') on the road from Maiduguri to Baga, a market port on the lake shore. He also provided the necessary recommendations to senior officials of the Ministry of Works, Maiduguri, who were in charge of the buildings and generously made them available to us, rent free, for the duration of the expedition *(Plate 2.4-6)*.

The main house was built of wooden slats and plywood on a concrete foundation, with a corrugated aluminium roof which was tethered to the ground by means of heavy twisted wires embedded at the bottom in concrete bases. We soon saw the wisdom of this arrangement: when a storm or tornado began to blow up from the north-east, a howling, sand-laden gale tore across the land from the lake, so that the whole house strained at its tethers and foundations in an eager attempt to become airborne. An abundant and constant piped supply of warm or very hot artesian water, which turned a rusty brown colour on contact with the air, gushed from the taps. Our bath was thus stained brown in colour. A gas cooker, kerosene fridge, and standard government furniture amply met our needs, and indeed added a touch of very real comfort to the expedition's amenities. When we took up residence at Minetti, the new road from Maiduguri to Baga (with a branch also to Kukawa), had not been long completed.

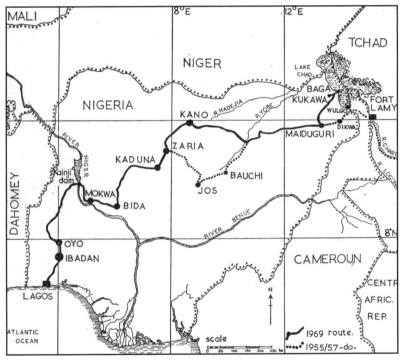

Map 4. The overland route across Nigeria taken by the 1969/70 expedition.

It was built on a raised foundation rising to a causeway as it crossed the inundation flats towards Baga. Here it was solidly built with stone sides, marked by white bollards ready for some future occasion when the lake might stage another maximal rise in level. However, the road engineers had been requested to build sand ramps up to the causeway at intervals to allow livestock to cross. Soon after they had completed these, some elephants were seen to be using the new crossing, and the local people thereafter referred to them as the 'elephant ramps'. The tarmac of this road is double width, with a wide sandy verge which is used in many places by donkeys, horses and camels as the tarmac becomes very hot and is hard on the hooves.

Huge diesel lorries, some with enormous trailers, hurtle along this road - a far cry from the old track through the sand - carrying lake produce south: tons of dried fish, a little frozen fish, grass mats, indigo, natron (imported from Niger and Tchad Republics), wheat (from Malamfatori on the Niger Republic border), and some guinea corn. There is also quite an important trade in thatching grass and firewood for Maiduguri, as almost all Nigerians still cook on wood fires and many live in thatched huts. Passenger vehicles of standard design, namely a wooden body locally built on to a lorry chassis so as to provide the maximum possible carrying capacity, were always crammed solid with people, sheep and goats, chickens, and odd-shaped bundles of luggage. The freight lorries also carried a large complement of passengers on top of the load, and were usually terrifyingly top heavy. These vehicles hurtled the 200 km (126 miles) from Baga to Maiduguri and back at great speed, the driver and his mate often changing places while making speeds of about 100 km.p.h. It was therefore not entirely surprising that there were two fatal crashes during the very first week that I visited Baga, Minetti and Kukawa.

Our neighbours at Minetti were the Kanuri people of Tasha. 'Tasha' is an adaptation of the word 'station', and here meant 'bus station'. Here a small village had arisen in the buildings formerly used by the road workers, and it served both as a bus stop and a pick-up point for local produce going to and from Maiduguri market. There were also points along this road where villagers, denied the privilege of a road to their own doorstep, came in on horse, camel, or donkey and 'parked' them for the day while they travelled on to Maiduguri by bus or lorry, collecting them again in the evening. Both Richard and Mohammedu used these buses and lorries on occasion with various degrees of discomfort, the main trouble being their complete irregularity. It cost N10/-* for a front seat and N6/- for a rear seat for the single journey of ninety miles between Minetti and Maiduguri *(See: Plate 21.1)*.

---

* Nigerian shillings, later superceded by the Naira

## Chapter 2 - A yacht on Lake Chad

Our other immediate neighbours were Fulani herdsmen and women who, together with our Kanuri neighbours, brought their livestock to the artesian overspill to drink. They also supplied us with fresh goat's and cow's milk. Their donkeys brayed around us, their horses foaled, and their hobbled camels browsed in the bush nearby. It was a friendly community which accepted us kindly and protectively *(See: Plate 23.1 and 5).*

It was the District Head of Kukawa who first suggested that we should launch the yacht at Portofino *(Map 5, page 83/84).* It goes without saying that this was another road camp built and named by the Italian road engineers, but unlike Minetti it was situated at the lake shore and even had a little sandy beach. It was much cleaner than the Baga foreshore, and had deeper water near at hand. We had discussed the question of how to launch the yacht with a number of supposedly knowledgeable people. Some said it would be quite impossible anywhere other than at Fort Lamy on the Chari River; others said we should have to dig a channel first or build a mile-long jetty. Others suggested the use of oxen to pull the boat on its trailer out to deep water. The District Head, however, took one look at the yacht, and said that twenty fishermen at Portofino could pull it out by hand on its trailer, and moreover he kindly offered to organise the operation for us as a community effort *(Plate 2.1).*

We therefore went out in the *Sportyak* to determine the channel with the firmest and most mud-free bottom leading soonest to deep water and a suitable anchorage. Then on 18 August 1969 we towed the *Jolly Hippo* to Portofino, right down to the water's edge, disconnected the trailer from the car, tied on long transverse 4cm pipes, and two long ropes, and handed over to the fishermen and their children. We expected to hear groans and yells indicative of a tremendous output of effort, but, in fact, amid laughter and good humoured chatter, they pulled the yacht and trailer straight out into the water as if it had been a baby's buggy. Within only a few minutes Richard yelled: 'She's floating', and indeed she was gracefully drifting off her chocks on the trailer as if she wondered what all the fuss was about.

But both we ourselves and the *Jolly Hippo* still had much to learn about Lake Chad and its moody tantrums, for, only a few days later, after a night gale, she was washed aground on a sandbank. We hastened to the shore, straining our eyes to locate the aluminium mast and blue hull. The Village Head approached us, rather accusingly, we thought. 'twenty men,' he declared, 'had struggled from dawn to re-float her, but had completely failed.' We asked why they had not sent to tell us, but apparently this thought had not occurred to them.

She lay far west of her anchorage, listing so badly that we feared she had been damaged. Drew Barlow (who was staying with us for a while), Richard and Mohammedu waded out to her. They rocked, pulled and pushed, but to no avail. Meanwhile a north-east wind was blowing up, and a mighty arch of heavy cloud was advancing towards us across the lowering sky. It looked and felt ominous. Far out across the lagoon, I saw the water whipped into choppy waves, the grass and reeds bending to the growing force of the wind. Fishermen out there jumped from their canoes and could be seen pushing them hastily deep into the reed islands, as the wind whistled through the grasses. Camels lay down on the sand, donkeys turned their hindquarters to the storm and hung their heads, and people ran to their huts.

I saw the three men leave the *Jolly Hippo* and start hurrying through the lake water towards the car, emerging shivering with cold. One fisherman, as he pulled his canoe higher up the beach, remarked casually: 'In less than two hours your boat will be floating.' I was puzzled: how, and why? As the cloud advanced the gale tore at the huts, and everything around us, anything which was not secured flying madly away. The world was first a cold greyish-white, then navy blue, and finally almost black as the cloud opened and spewed forth sheets of vicious, pelting rain. *(Plate 6.1-2)*.

It was then that we saw the lake advancing up the shore towards us as we now sat in the car - one metre, two metres, ten metres. Canoes which had been drawn well up the shore were floating, and the prefabricated roof of a hut which was under construction began bobbing up and down. In some places the water marched up fifty metres, and in one place about one hundred metres. The level must have risen fully 30cm, and, as the rain eased, and the yacht once more became visible, we saw that she was indeed floating again. As quickly as it had arisen, the storm subsided, and soon Drew, Richard and Mohammedu were out there again at the *Jolly Hippo's* side, gently rocking her to ease the keel from the sand, then towing her forward to her former anchorage. One thing was quite clear to us now: a secure mooring was essential.

Soundings were taken to locate the deepest, safest place. The place selected was somewhat exposed, but being in the lee of two low-lying grass islands, we hoped that these might provide at least some protection from vagrant floating islands and north-easterly gales, a supposition which I was later to disprove in most irritating circumstances. We made two concrete blocks, moulding them in empty 18 litre kerosene tins, and embedding in each the enormous U-bolt of a lorry spring to serve as an eyelet.

These we chained together with a one-metre length of galvanised chain, and prepared to drop them at the mooring point. How to drop them was quite a problem, but eventually we decided that Richard should drop one from the corner of a Sportyak while Mohammedu and I simultaneously dropped the other from the higher and more awkward deck of the yacht. What we had overlooked was the fact that the combined weight of Richard plus concrete block would inevitably overturn the *Sportyak*.

On the word, we dropped our heavy blocks, and at the same instant the Sportyak turned turtle, throwing Richard in the air, to land on his feet most gracefully on its upturned bottom. This concrete mooring stood all further tests of gale and storm, and the Jolly Hippo rode comfortably at anchor there, whenever in her home port, for the remainder of the expedition. We also planted an indicator post nearby so that we could keep watch on the level of the lake, and attached a marker buoy, actually an empty cooking gas cylinder, to the mooring.

We now turned to the completion of on-board preparation of the yacht, such as mounting and wiring the new prow spotlight, fitting removable mosquito net frames to the cabin hatches, improving the galley, swinging the compass and making a mount for the log. The inshore graphic echo sounder also had to be mounted. As the lake is so shallow, the makers had advised us to mount the transducer on a removable arm, rather than faired-in to the underside of the hull. With the limited equipment available, this proved to be a surprisingly awkward task, but eventually the instrument was satisfactorily fitted, and produced most interesting traces of the topography of the lake floor *(See Diagram 4, pages 96 - 98)*.

We attempted our maiden run during another visit by Drew Barlow, this time accompanied by his father, Lieutenant-Colonel H. E. Barlow. Unfortunately we kept running into shallows and mud-banks, and getting the yacht stuck. (We had yet to discover that it could always be readily freed again if the helmsman immediately sang out for all passengers and crew to move to one side and lean out over the water, while he put the helm hard over with the engine running at full speed ahead.) The day was saved, however, when a herd of elephants was spotted, standing 'in' - or should one say 'through'? - a floating island about 2.4km north of where we had again become firmly embedded in the mud - so we all left the *Jolly Hippo...* which then at once floated free again! Then we set off in a *Sportyak,* using the outboard engine. Unfortunately, as we got near the floating island, we encountered saturated mud and could not approach closer than about 135m. Nevertheless we took some photographs and I had a good opportunity carefully to study these elephants in their favoured habitat with binoculars.

The herd consisted of eight very large, middle-aged bulls, several bearing good medium-sized (about 25-30kg each side) ivory. Two had lost the tips of their tails, one apparently had ear ache (he kept feeling the orifice with his trunk tip), one had an anal hernia (the usual anal flap being greatly distended), one had an open injury on his flank, and one had a broken ear, the upper part having been cut right through for a distance of about 15cm to 20cm. All were nevertheless fat and mobile with supple hide and a good stance. As time went on we came to know the Chad elephants well and I was able to make a more

## Chapter 2 - A yacht on Lake Chad

detailed and precise study of three of them *(See Chapter 6, The Fauna).*

We now made a number of short trial voyages in the vicinity of Portofino and Baga in order to test the usefulness of our maps in terms of the current level of the lake and disposition of the floating islands, and how best, in the circumstances, to navigate.

Navigation by dead reckoning was the only practical method - using only the ship's compass and log - but this had to be supplemented by the simple expedient of asking the names of any convenient fixed islands whenever we encountered fishermen, and then trying to locate these on the maps. As the names spoken and those written on the maps did not always seem to tally, there was ample scope for imaginative interpretation.

Before embarking on the expedition, I was asked by a number of keen sailors what sort of landmarks we should have at Lake Chad as the shore line is very low and featureless. I was to discover that each of the towns of the main shoreline, and some dune island villages, contain some trees, and these, as well as the huts, become so distorted in the vertical plane during the hotter part of the day, that they serve as recognisable landmarks. *(See Plate 3.4).* At night, however, it is extremely difficult to distinguish between one island and another, or between island and shoreline. On one voyage my night navigation broke down dismally when I thoughtlessly used a hurricane lamp to illuminate the marine compass as I steered: I realised my *faux pas* just in time to calculate the error and adjust our course before running into a particularly shallow and treacherous area. However, the mistake cost us several hours' cruising and some very unpleasant encounters with nylon fish nets. At the end of these early trial voyages it also came to light that the original swinging of the compass together with the entries on our table of deviations had embodied major errors so that all the early navigation was confused, puzzling and inaccurate. Happily I was soon able to rectify these errors, and from December onwards navigation on Lake Chad proved to be surprisingly simple.

Two major horrors to be encountered on Lake Chad nowadays by the crews of powered vessels are, first, nylon gill-nets, and, secondly, nylon lines loaded with fish-hooks. Our first experience of these was salutary. We had made a day's run around our local floating islands and were returning home to Portofino, now about three miles run to the west.

We were heading into the setting sun, when suddenly the engine gulped, hesitated and failed. Richard jumped down into the water to discover that a nylon gill-net was not only entwined around our propeller shaft, but had 'welded' itself into a solid nylon ring. The water was just shallow enough for each of us to stand, our legs, from the knees down, embedded in that thick black oozing mud, and our heads just awash at the stern of the yacht. Reaching under the stern with a knife and scissors, we could just get at the nylon tangle and ring. Mohammedu and Richard also tried diving underneath but achieved little that way, and merely got exhausted.

It took us two hours to cut, rub, tear and hew that nylon free. Then we drank some hot soup, started the engine, and got going again. The moon had risen now, throwing a gleaming path across the water. It was absolutely beautiful, and for the moment so peaceful . . .

We had only gone another three or four hundred metres when the engine again gulped, hesitated and failed. We stared into the black water behind the boat in blank and heavy silence. During our first encounter with the nylon net, a fisherman had come alongside in his canoe, saying it was his net. He was obviously sorry for us, and cleared the main body of the net away. We were concentrating so hard on clearing our propeller shaft that not one of us had noticed that when he took up his net, he only carried it a few hundred metres away and then carefully relaid it!

He now came alongside again and helped us for a while apparently in no way resentful of the damage done twice over to his net. He still seemed sympathetic, but this time we remembered to ask him not to re-lay it in our immediate homeward path. It was midnight before we found our mooring in Portofino harbour and wearily rowed ashore in the dinghy.

From then on one member of the crew was always on watch for *raga* (fish net). These nets may be anything from 20 metres to 1½kms in length, weighted with the old bones of oxen and horses (there are no rocks or stones in or near Lake Chad)* and upheld by floats of cut papyrus stems, each about one metre in length. At each end a papyrus frond, sometimes complete with its 'topknot', or a pole or double pole, is stuck vertically into the lake bottom. If there is any breeze at all, the waves and wavelets make it intensely difficult to spot these floats and markers. As time went on, however, both Mohammedu and I developed such a severe neurosis about running into these nets that we seemed to 'feel' their presence long before we actually saw them.

The other horror is the nylon fish line, loaded with innumerable hooks. These may or may not be baited and are usually suspended between a vertical marker-frond or pole and an island. The hooks dangle on short lengths of nylon from 30-45cm long, at intervals of about 30cm, throughout the entire length of the line. As these lines are not usually upheld by floats, they are less readily detected than the nets. When these get entwined around the shaft, it is even more difficult to release them than the nets because the hooks also become 'welded' into the nylon ring that forms.

---

* Except for the great Hadjir el Hamis rocks near the mouth of the Chari River *(See Plate 27:2)*

Dr Carling, a medical missionary working on Lake Chad, recounted that he had suffered similar experiences, and on one occasion had a nylon net float back around him as he worked to free the propeller of the mission launch: a most alarming situation. The mission launch, however, was much more powerful than the Jolly Hippo, and could generally cut its way through unseen nets without entanglement.

The new research vessel, imported to Tchad Republic in 1969 for use on Lake Chad, by the Office de la Recherche Scientifique et Technique Outre-Mer (O.R.S.T.O.M.), a French organisation, is fitted with two propellers which can be raised from the water for cleaning and repair, and, presumably, easy removal of entangled nets and lines. It would have been quite simple to have fitted a net-deflector to the hull of our yacht before leaving England, had we realised the need.

Unfortunately I had not been warned about this hazard and could not have anticipated it, as the introduction and extensive use of these nylon nets in Lake Chad is rather recent. Naturally, the nets and lines never troubled us at all when we were operating only under sail. During our early attempts at sailing, we were becalmed from about noon to 1.30 or 2.00 p.m. every day, and the November breezes were usually too light for our immediate purposes.

At that time, therefore, we mainly cruised under power during our trial runs, only unfurling the genoa to provide boost if we were in a hurry. We could then make about six knots.

I find it very surprising that, since Denham's visit to Lake Chad almost 150 years ago, the introduction of powered boats has been essentially sporadic and individual. The use of sail has been negligible and, as far as I could discover, ours was the first true sloop to sail Chadian waters. About the turn of the century the French imported two steam boats to control piracy on the lake, and later the British Resident at Maiduguri had two huge canoes imported overland from Nupe country in southern Nigeria for administrative and police purposes.

Captain Smyth, a British District Officer, tried to introduce the use of sails to the Yedina fishermen, but without success. Fisheries and Customs control officers introduced large wooden canoes (built of planks and fitted with outboard motors) and steel barges (built in Fort Lamy for use on the Chari River) fitted with powerful inboard or outboard motors, both of which have from time to time been used as freight vessels on the Lake. Both the Federal Fisheries Research Station at Malamfatori in Nigeria and O.R.S.T.O.M. in Fort Lamy several years ago introduced diesel-powered research launches suitable for use on the lake. The new, larger research vessel, imported by O.R.S.T. O.M. in 1969, is (up to the time of writing, 1971) the biggest ever launched on Lake Chad, being l4m long, equipped with laboratories, cabins, decks, and the most up-to-date research equipment. In 1964 the Sudan United Mission imported a 10m cruiser, *Albishir,* for medical work on the lake, and is (1970) about to launch a new 15m hospital cruiser, *Albarka,* complete with an operating theatre . *

In addition to the few powered boats in operation on Lake Chad, the Mission Aviation Fellowship in 1966 introduced a light amphibian plane to aid medical missionary work and in late 1969 a British firm brought a hovercraft for three days' trials on the lake. Casual sightings of powered boats as one sails or cruises among the islands are thus still few and far between.

**As yet (1970) the jarring sounds of modern technology rarely intrude to disturb the mystique of the mirage-lake.**

---

* This vessel is now (1970) fully in service on Lake Chad, but see ch. 12, eventually destroyed by rebels

# Chapter 3

# Lagoons, dunes and floating islands

Through December 1969 and January to March 1970, Mohammedu and I came to know the lake better and better. Usually assisted by a Yedina guide, as interpreter, we cruised into the northern and eastern archipelagos, as well as among the islands and the lagoons fringing the southern shores. On many occasions we left the *Jolly Hippo* and took to fishermen's canoes of various types. In the early days we soon became very weary and stiff with sitting motionless for long periods on the hard thwarts of these canoes. These were really just narrow struts that kept the canoe in shape. But later on we became so accustomed to them that we sometimes endured up to six hours' 'incarceration' at a stretch.

Only once did we find ourselves in a positively unsafe canoe, and then we insisted on our immediate return to land and re-embarkation in a vessel with higher gunwales. In high winds canoes are sometimes swamped or overturn. The first Yedina fisherman to pole us into the lagoons near Portofino in October was sadly drowned in December when his boat capsized in a storm on the open waters. One of his fellow-boatmen was a strong swimmer and swam the 8.5km back to land.

The most trying aspect of long canoe trips was our prolonged exposure to the blazing sun. From 7.00 a.m. to 4.30 p.m. it beat down on us mercilessly, the added reflection from the water enhancing its burning effect on the skin. Although I never sunbathed in the accepted sense of the word, and always wore a strong cotton shirt, the skin of my back became deeply bronzed. We nearly always found it necessary to wear hats.

When Miss Mary Thomas visited us from England, she calmly adopted the local fashion by wearing the large, wide-brimmed, conical-crowned straw hat normally worn by the Fulani cattle men. This acted as a most efficient and convenient sun hat, parasol and (in rainy weather) umbrella, and was obviously the ideal hat for lake wear at all times except during a gale.

There was usually some breeze on the lake except between noon and about 1.30 to 2.00 p.m. as well as for part of the night, but within the lagoons it was often very still and sultry, the heat seeming intense and pressurised. It was impossible to take the *Jolly Hippo* into many of the in-shore lagoons because frequently they were too shallow. Also some of those on the western shores contained the submerged, or partly submerged, standing lower portions of trunks of dead trees that must, at some time in the past (possibly during the long period of lake desiccation in the early part of the twentieth century), have comprised savannah woodland. It was on these occasions that the *Sportyak* dinghies, with or without the outboard motors, came into their own.

Only local canoes, however, could tackle a passage actually through or across a floating island. The boatman would then force his way along hippo or elephant trails within the papyrus and *Phragmites* reed grass. Immediately we were enclosed within the floating vegetation, the humid breathless heat seemed to envelop us, and hordes of biting mosquitoes, and gnats, would dive-bomb the exposed parts of our heads and arms, gorging themselves viciously on our blood. Meanwhile hundreds of little red biting ants would swarm on to the canoe, sinking their hungry jaws into our sunburnt skin, cheered on by dozens of jolly little weaver birds nesting in the papyrus.

As long as the engine of the *Jolly Hippo* worked healthily, cruising was a simple and pleasurable way of exploring the shallower lagoons and more crowded islands. We would stock the galley with a week's supply of food and kerosene, plot our proposed course, and head out across Portofino Bay with its submerged grass and floating islands until we reached the open waters when we could cut the engine and set the sails,

Beyond Portofino Bay and Baga Headland, we usually encountered more interesting conditions with stronger

## Chapter 3 - Lagoons, dunes and floating islands

winds and sizeable waves. *The Jolly Hippo* was designed as an ocean-going yacht and rode the waves to the manner born. In a 60km.p.h. (force 8 Beaufort scale) wind, she used to remind me of a show-jumping horse, and one rode her in just the same way, relaxed and confident in her complete suitability and reliability. I think the only times when I experienced any anxiety about the safety of the yacht were when the waves were running at between 1.8m and 2.4m high and the echo sounder was simultaneously showing a very uneven shallow floor, studded with fossilised dunes (or 'bench islands') whose summits were only just submerged. Then I did very seriously wonder just how the vessel would stand up to the shock if a wave raised her to its own summit and then dropped her on to the head of a temporarily exposed underwater dune. I am thankful that this eventuality never materialised.

As the evening drew in, we would seek shelter close in under the lee of a dune or attached floating island, and drop anchor, hastening through our evening meal to avoid the winged hordes, the myriads of mosquitoes, that would invade the boat. Then we slapped the net frames over the cabin hatches and ran up the nets for those sleeping on deck, diving inside as quickly as possible.

Slipping between the islands in the hazy heat, pitching and rolling across the open stretches, and gently rocking at anchor under the stars were timeless experiences. We would watch the fish nosing the surface of the water into ever-widening circles, or a snake cutting a V-shaped trail between two islands; we would enjoy the chorus of the tree frogs and crickets, and the squeak of the bats, while the night sky hung overhead, a velvety blue curtain pierced by a million little lights. And then would boom out across the water the guffaw of a bull hippo. 'Ugh; huh-huh-huh! Ugh; huh-huh-huh!'

We took it in turns at the helm. At first Mohammedu was unbelieving about the efficacy of dead reckoning, but he soon became convinced and learned to steer a course by the compass. Galley duties were shared, although Mohammedu really preferred to cook his own kind of food separately. But whatever the number, whether two or more, we usually ate together and had the same food whenever we all slept aboard. At other times one or two members of the party would sleep and eat ashore in a village or a floating camp, and always we found the people hospitable and friendly, welcoming us with gifts and warm greetings. It was easy, in the dry weather, to sleep in the villages and camps, all that was necessary being to set up one's camp bed, or, in Mohammedu's case, unroll his sleeping mat, hang the mosquito net (we used the 'bell-tent' type) from a tree, pole or hut ridge, and settle down.

On one occasion early in the expedition, I slept alone aboard the yacht, while the other members of the crew slept ashore on a dune island in the midst of a rather complex archipelago in the northern part of the lake.

In the morning I awoke to find that the anchor had not embedded itself properly in the lake floor, and the yacht had drifted far out on the open water. Fortunately the north-easterly breeze had remained steady and by heading directly into it, I was relieved to find myself back at the correct island. Had the breeze veered, I might well have had difficulty in re-locating my position and in recognising the correct island, as many of the islands looked rather similar and there were few inhabitants in this area whom we might ask.

It was during the visit of Mary Thomas from England that the engine of the *Jolly Hippo* first seriously broke down. It was very regrettable that, in the early stages of the Expedition, a volunteer who undertook the maintenance of the engine neglected to maintain the oil levels correctly or to keep up the prescribed maintenance schedule detailed in the makers' service manual. This failure was to prove very costly to the expedition both as to our subsequent mobility on the lake, and as to the replacement of parts and repairs. In spite of many further changes of oil, and attempts to clear the lines of congealed sludge, the oil never again circulated freely until the shear pins of the shaft eventually sheared and the entire engine had perforce to be removed.

Mary only had a week of her vacation left, and we were heading towards Baga Sola to rendezvous with the amphibian plane belonging to the Mission Aviation

Fellowship, based at Fort Lamy in Tchad Republic. We had chartered it with the intention of flying over the north-eastern islands and lake littoral in the hope of seeing, and perhaps counting, elephant, hippo and any of the larger antelopes that might be visible.

It was a perfect day as we set out, and we were delighted with the silky, peaceful beauty of the lake. We had completed 32 nautical miles, having successfully circumnavigated a number of complex groups of floating islands, and now it was late afternoon. Only about seven nautical miles remained until we should reach Baga Sola, the Customs Port on the Tchadian shore. We were neatly on course, doing about 5 knots, when, without any warning at all, the engine screamed, and the yacht just stopped cruising, although the engine was still running.

We were near an uninhabited floating island, and about a 1.5km ahead lay a long dune island on which we could see cattle, palm trees and a village. We dropped anchor immediately in about 2.5 metres of water. First Mohammedu, and then I, dived under the stern to investigate but could find nothing wrong there. Then I crawled into the unbelievably cramped and confined engine compartment and tinkered with the engine. The gear box was very hot and I suspected that it had overheated due to oil circulation failure and that maybe some part had broken inside. If so, there was little I could do to repair it unaided.

Nevertheless I dived under the boat once more, just to make sure nothing was broken or clogging the propeller. While under water I had a strange feeling that I was not alone down there, so, when I surfaced, I was not entirely taken by surprise when my eyes met those of a hippo glaring at me from about 18m away! He looked so offended and aggressive that I wasted no time at all getting aboard again to ponder the situation in greater safety *(Diag.5, page 100).*

It was most unfortunate that, in order to minimise baggage, we had left the outboard motor for the dinghy at Minetti. I therefore saw no alternative to towing the yacht to the dune island with the dinghy by rowing. We were far too exposed in our present position, and yet there was insufficient breeze to sail. So we set off, Mary calling my stroke, pointing out the direction, and cheering me on by playing pop music on the radio, while Mohammedu covered his head with a sack and shivered.

We found a fairly firm landing place, free of floating fringe vegetation, on the dune island, which we now identified as Samiya. We drew close in-shore and dropped anchor. Mary cooked the supper. Mohammedu prayed on the fore-deck and I had another go at the engine.

As darkness fell we ate our supper together somewhat meditatively on the deck, only to be rudely interrupted and galvanised into action by the arrival of a very angry posse of hippos.

## PLATE 1
## MY 1955 VISIT TO LAKE CHAD.

1. *(left).* My first sight of papyrus canoes.

2. *(below).* A Yedina fisherman poles Mohammedu and Edith Jackson across a shallow lagoon towards the open lake. A low line of floating vegetation fringes the horizon

3. *(above).* Traditional compound in Dikwa, once the military headquarters, in the 1880s- 90s, of the brigand Rabeh showing the sturdy dried mud walls of houses with flat roofs and gutters.
A Shuwa Arab family drive their laden pack ox to market.
4. *(above right).* Poling across fields flooded by the lake's inundation near Gambaru.

# PLATE 2
## A YACHT, THE *JOLLY HIPPO,* ON LAKE CHAD
## 1969 - 1970.

1. *(above left.)* Our arrival with the *Jolly Hippo* at Lake Chad.
2. *(above right).* The *Sportyak* dinghy which proved to be tough and suitable.

3. *(above left)* Yedina fishermen in their *kadais* crowd around the *Jolly Hippo* out on the lake. 4 and 5. *(above right).* Mohammedu and his wife Binta stayed at the lake with me for the whole eight months of the expedition.

6. *(above left).* Our base camp, the former road-builder's compound at Mile Ninety on the Maiduguri/Baga road, known locally as *Minetti*.
7. *(above right).* Our splendid *Sportyak* dinghy with jet outboard.

## PLATE 3
## LAGOONS, DUNES and FLOATING ISLANDS (A)

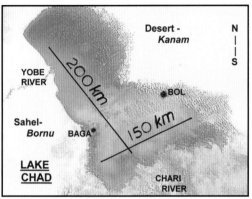

1. *(left).* Lake Chad - a 'desert of water' with about 200km NNW/SSE and 150km SSW/NNE of open water navigable in the *Jolly Hippo* yacht in January 1970.
2. *(below left).* Sky and water merge: evening.
3. *(below right). Phragmites australis* flowering heads in the dawn sunlight

4. *(above).* Floating island of *Phragmites* in windy weather

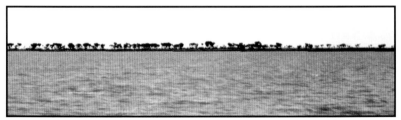

5. *(above).* Navigation recognition points might be just shimmering silhouettes of trees on dune islands!

# PLATE 4
## LAGOONS, DUNES AND FLOATING ISLANDS (B)

1. *(right).* Emerging between floating islands, from a lagoon on to the open waters of the lake

2. *(left)* Inside a floating island of papyrus. Papyrus is the basic resource for making Yedina boats *(kadais)* and huts.

3. *(above).* Broken papyrus stem, triangular in cross-section.

4. *(above).* Papyrus head.

These had approached their favoured landing place for their evening graze, and were apparently shocked to find it occupied by that intruder, the *Jolly Hippo*. This was too much for the bull, which now advanced to within a few yards of us, opened his maw and said in his deepest base voice: 'Ugh; huh-huh-huh ! Ugh; huh-huh-huh-huh!' I hastily grabbed the rifle, and shone a torch on him, at which he tactfully withdrew, and although they sloshed past, between us and the bank, later in the night, they did not really worry us again.

Next morning we were up and had finished breakfast before the dawn broke. I hoisted the red foresail and let it flap in the hope that it might attract the attention of the pilot of the 'plane in case he should try to look for us. Mohammedu set off across the island to contact the villagers, I went down again to contemplate the uncooperative engine, and Mary kept watch on deck.

At 7.05a.m. we heard the 'plane, and watched it fly in towards Baga Sola, and disappear behind the palm trees. What would the pilot do, we wondered. Would he search for us? The real crux of our anxiety was not that we were marooned on a genuine desert island, complete with sand and palm trees, but that Mary's air passage in six days' time from Kano to England was now at stake. Not only had we no means of communication whereby to cancel or postpone her flight, but if our present pilot failed to see us,

*Chapter 3 - Lagoons, dunes and floating islands*

the only two means left open to us of returning to Portofino - namely sailing or canoeing - were too slow and uncertain to enable us to arrive in time to drive the 770kms to Kano and calmly walk on to the tarmac the prescribed hour before take-off time. So, weighing the alternatives, we now eagerly awaited the next turn of events.

About forty minutes later, we saw the 'plane take off in the distance and, to our surprise, head due north. I sighed and squeezed myself back into the engine compartment. Suddenly Mary yelled 'It's coming; it's coming here!' Extricating myself so vigorously from the engine room that all my tools fell into the bilge, I dashed up on deck in time to see the 'plane skim across the water, and stop some 30m from us. We soon hauled it alongside and moored it.

The pilot had brought three unscheduled passengers with him, intending to fly them on to an island called Haikalou after completing the chartered flight for us. When he failed to find us at the rendezvous he assumed we were not coming and decided to fly straight on to Haikalou, but one of his passengers, using a telephoto lens on his camera, spotted our red sail flapping, and they came down to look.

Our kind visitors now took control of the situation, and one of them, Mr Walter Utermann, being of a mechanical turn of skill, decided to take out the entire engine. It was no mean task.

Eventually, with everyone helping, it was hauled up on ropes and the trouble identified. The gearbox oil had failed to circulate, and had congealed into a fearsome black glue, while the shear pins of the shaft were shattered.

We now held a council: since there were no boat yards on Lake Chad except for one at Fort Lamy, the only sensible course seemed to be to abandon the yacht where it was. The pilot would fly us to the Nigerian boundary and land us near some canoes. The fishermen would be asked to take us on to Portofino. He would then return and fly the visitors to Haikalou, and finally fly the gearbox of the yacht to Fort Lamy.

Unfortunately, the pilot was not permitted to fly over Nigerian territory at that time as the Nigerian Civil War was still in progress and all flying was restricted. We wondered how safe the yacht would be, lying there at the mercy of the former 'pirates of the papyrus', but could see no alternative other than to leave it. So we gathered together as many of our more valuable items of equipment and baggage as the 'plane was permitted to carry and clambered aboard, leaving our visitors to tidy up the *Jolly Hippo* and reinforce her moorings.

The little 'plane had a struggle disentangling its floats from the water due to its maximal load, but eventually we were airborne, heading swiftly back towards the Nigerian border. Near the international boundary we spotted two fishing canoes, so our pilot skimmed down and across the water, stopping gently just beside them. To say that they looked astonished understates the situation, and when Mary, Mohammedu and I clambered down on to the floats their eyes really did nearly pop out of their heads. 'They are *our* people', they said! Indeed, the fishermen in both canoes were from Portofino village itself and were well known to us.

Although their canoes were heavily laden with a splendid catch of fish, they took us aboard, Mohammedu in one, clutching the radio and the rifles, with one camera slung round his neck, and the bedrolls at his feet, and Mary and me in the other, slung around with more cameras, binoculars, the folder of maps, log book and notes, and the picnic box. We settled down atop the huge piles of fish, many of which were still wriggling, and waved the pilot off as he sped away right over our heads.

It was several kilometres' run into Portofino but the boatmen hoisted a blanket as a sail in each canoe and we ran in before the breeze at a spanking 4 or 5 knots, the water creaming away under the sharp bows. It was all terrific fun, and the Portofino villagers soon spotted us, and ran down to the beach to receive us, kindly hands quickly carrying the baggage to the car, while Mohammedu was hard pressed to answer all their questions.

Mary caught her plane home to England, very sad that

she could not stay longer to enjoy more real-life adventure, Chad style. She went so far as to admit that she would happily have forfeited the air passage home for the sake of a longer stay on our desert island, if only in order to sample a diet of hippo meat and water lettuce.

Mohammedu and I now concentrated on the lagoons fringing the southern part of the lake along the Nigerian shore, mainly using local canoes for longer excursions. It was here that we saw a canoe sink in rough weather and a high wind, its over-numerous occupants wading ashore disconsolate through some 1.5m depth of water in damp disarray. It was here that we saw how high winds set up strong currents, driving the water fiercely through bottlenecks between the vagrant floating islands and anchored papyrus rafts. And it was here that we became well acquainted with the Lake Chad elephants, monitor lizards, otters and terrapins.

In February word reached me that Dr Carling of the Sudan United Mission would soon be making a medical visit to Tchadian waters and could take me to collect the *Jolly Hippo.* So Mohammedu and I boarded the Mission launch, *Albishir,* travelling as passengers. The accommodation seemed most spacious after that of the *Jolly Hippo.* Sitting on the upper deck of the 10m Albishir gave one a different and less intimate view of the lake than that obtained aboard the *Jolly Hippo.* The bigger cruiser made about 7 or 8 knots and was less affected by muddy or weedy patches and fish nets. Dr Carling knows Lake Chad so well that he seems to navigate more by 'feel' than by his compass, and it was with no small sense of embarrassment that I discovered that I had apparently led him astray at one point as a result of quasi-mathematical reckoning derived from supposed speed (he did not use a log), time taken and compass bearing. However, even he admitted to some doubt as to our whereabouts.

In spite of this setback (which cost us about an hour or two) we did eventually arrive at the *Jolly Hippo's* mooring place. But she was not there. The doctor had been very patient over my navigational error, but he was now too intensely polite for my comfort. It was quite clear, although he did not actually say so, that he felt (justifiably) that, to cap everything, I had now mistaken the place. His friendliness grew alarmingly cool after we had called out to the boatmen of no less than two separate fishing boats, and been told that they 'had never seen such a boat', and, 'No! absolutely no such boat had ever been tied up at that place on this island'.

But Mohammedu and I were both insistent: this was indeed the very place, and there was the village and the palm trees - 'And,' I added, 'the cows were over there just where they are now.' The doctor's tone was ice-cold when he said: 'I suppose you have considered the possibility that these may be different cows?' Nevertheless, chilly as was the conversational atmosphere, he did offer to row ashore with Mohammedu and call on the villagers. I hastily looked in my notebook and said I had written down the name of the village: the villagers called it *Kambe* and we thought this was also called *Samiya* as on our map, so he could use this as a check on my veracity if he liked! Thrusting a 5-shilling note into Mohammedu's hand I whispered 'You'd better *buy* the truth, if necessary.' They were over at the island a long time talking to the villagers, but eventually I saw Mohammedu hand over the money, and a few minutes later they returned.

They had at last learned that a large white motor boat had come and towed away the *Jolly Hippo*. This was all the information that could be bought with the five shillings. Whose boat? Which way did it go? Which day? We were left to speculate uneasily as to the answers to these questions. Puzzling over what to do next, the doctor suggested going to the Customs Post at Baga Sola (no less than two hours' cruising time to go there and back), to see if they could shed any light on the mystery.

Baga Sola was a derelict-looking place. A few sand-coloured buildings, a few shade trees, a few half-submerged, waterlogged boats, and a few camels kneeling beside their saddles and loads of 'potash' (natron), munching drily. The Tchadian Customs officials were sorry: no, they had not seized the yacht and did not know who had done so; anyway, none of their boats was serviceable now. True, they had a radio, but it did not work now . . . awfully sorry . . . '*Au revoir. Madame! M'sieur!*'.

It was getting late, and the doctor was anxious to be on his way to Haikalou to his rendezvous with the 'plane at 8.00 a.m. next morning. We must press on now, at full steam ahead, into the night. When we reached a certain point, we would anchor and sleep for a few hours, and then go on again at dawn.

*Albishir* had a wonderful galley, stocked most adequately for the trip, meals being efficiently cooked for us by another missionary doctor who was also travelling with us as a passenger. After supper I sat on the upper deck pondering this intriguing lake. We were now passing through lagoons fringed with tall, lush floating papyrus. The air was very still, with not even the tiniest whisper of a breeze, just the swish of water against *Alhishir's* bows and the hum of her diesel engine.

The moon was full, casting deep unfathomable navy blue shadows. Bats whirred around the launch, and an owl hooted somewhere in the papyrus. The slap-slap of the fish jumping ahead of the launch jarred on the stillness of the night, seeming to emphasise the mystery and remoteness of this watery moonlit world. Some time, late that night, we anchored and slept, setting off again at dawn for the final run into Haikalou.

We had now left the maze of papyrus lagoons and were heading north among wide low dune islands on several of which large Yedina villages could be seen. Already the morning sun had begun to distort them, so that the huts looked grotesquely tall and clustered. Haikalou Island lay close to its neighbouring dune, and we found ourselves coming in through narrow straits of calm, shallow water. Scanning ahead with the binoculars I spotted a wooden jetty - some huts* - a gleaming mast - and then a blue and white hull. Snugly moored by the jetty just below the new-born mission Station at Haikalou lay the *Jolly Hippo.*

It was a fascinating moment to arrive at this place, for here was the moment of birth of this new mission station, a pioneer outreach of medical missionary work with all its promise as the means of bringing hope and healing to the people of this incredibly remote area. Earlier the mission plane had flown in a Swiss benefactor of this new outpost so that he might personally witness the actual birth of this little hospital. For his benefit, the Swiss flag flew gaily from a gleaming white flagstaff standing at the corner of the foundations of the hospital. Around it lay heaps of cement, sand, girders, gas cylinders, and all the paraphernalia of building construction work *(Plate 32.3).* The missionaries themselves were living in three round Yedina-style grass huts, and had made these comfortable and homely.

---

* The mission station and village were destroyed by rebels c. 1978

But there too was my yacht: how and why had it got there? Our kind friend Mr Walter Utermann had felt anxious about leaving it unguarded on Samiya island, so, when a motor barge from Fort Lamy had arrived bringing his building materials, he had arranged that the boatman should return to Samiya and bring the *Jolly Hippo* in tow to Haikalou. But his kindness had not ended there, for he had also re-fitted the repaired engine, and, apart from one item, it was all ready to move under its own power again. Parts which he could not properly replace, however, were the shear pins for the shaft, so he had made temporary replacements from nails.

I slept aboard that night, and Mohammedu and I set out for Portofino at dawn. It had been suggested that I should follow Dr Carling's return route in *Albashir,* as the 'plane was expected also to follow this route later in the morning on its way back to Fort Lamy. The suggestion seemed sensible, but when, later, Mr Utermann expressed doubts about the durability of the shear pins, I felt it wisest to follow a course where, in case of need, I could be sure of finding, sooner or later, sufficient wind to sail home. Unfortunately, both Dr Carling and the pilot had already left Haikalou by the time this decision was reached.

Once again it was a beautiful sparkling day, and as soon as we were clear of Haikalou and its neighbour island, I set a course almost due west towards the open water. We had no difficulty over the next fifteen nautical miles, except for occasional weedy patches overlying submerged fossil dunes. Some time during the morning we heard the 'plane flying south on its homeward journey and I fervently hoped that, when the pilot failed to see me, he would not waste time and fuel searching.

Later that morning, we were rounding another dune island when the engine laboured, the yacht slowed, there was an ominous 'click' in the shaft . . . and then we stopped. The shear pins had again sheared. It was dead calm and not a canoe was in sight. But this time I was under no obligation to arrive by any specified time at Portofino, and there were no 'planes, buses or trains to catch. This time I could sail home, take my time, and thoroughly enjoy it *(Diag.1, page 79)*.

We dropped anchor, cooked a good lunch and rested while waiting for a breeze. Then, true to form, about 1.30p.m., a soft breeze ruffled the water, and the yacht began to rock gently. At 2.00p.m. we unfurled the sails and made a long slow tack towards the north-west, hoping to clear the northern tip of the next dune island so as to be in position for a smart, close-hauled run towards the west.

As we drew level with the northern tip of the island, trying to pass it and get clear before turning, we found ourselves drawn across and back towards the shallows flanking the island, and only by going about rather desperately at the last moment did we avoid getting firmly grounded against the island in its sandy shallows. We actually made five unsuccessful attempts to round the island - an exercise that I found tiring but extremely interesting, for it is these curious wind patterns around the northern tips of the dune islands that account both for their existence and shape. The sixth time (I now felt that the previous efforts had provided sufficient educational data to satisfy my rapidly cooling curiosity), I first made a long north-easterly tack, so that my entire subsequent north-westerly manoeuvre would have no relation at all to the obstructive island, a move that carried us well clear and enabled us to make our westerly run as hoped during the evening.

From our maps we had little difficulty in identifying the dune islands of this area, as some of them had large, rather distinctive villages on them. As dusk fell, we finally cleared the last islands of this particular archipelago, and turned south with a good following breeze which kept us going until about 8.00p.m., when we were again becalmed.

We dropped anchor and prepared to have supper, when, as so often happened, a Yedina canoe came racing over towards us. The fishermen full of enquiries as to our welfare and full of curiosity about the *Jolly Hippo,* greeted and welcomed us. As usual we took them aboard and showed them around. Then they offered us a freshly caught Nile perch weighing 6.8kg, which we gratefully accepted, and, as soon as hey had left, prepared our supper. We slept until about midnight, when a breeze again sprang up. By now also the moon had risen, and we held steady on course due south until we ran into a complex maze of thick, papyrus islands and lagoons.

In the moonlight there seemed to be no way through, the papyrus just appearing as a dark rim all round the southern horizon and its flanks so that we seemed to be running into a huge cul-de-sac. However, as we kept on course, passages generally seemed to open up ahead of us. Each time I was forced to circumnavigate a floating island or barrier of papyrus, I quickly plotted the adjusted course on squared paper, regaining the original course as soon as a clear passage again opened up so we could recover our position.

Diagram 1. The Jolly Hippo, close-hauled during a run across open waters, with floating papyrus islands in the distance.

Due to the constant fragmentation of clumps and rafts of floating vegetation that break away from the anchored floating islands, we frequently encountered vagrant 'mini-islands' in our course, or, as now when running before the wind, accompanying us alongside. Most of these, however, travelled more slowly than the yacht under full sail, and we quickly outstripped them. Sometimes, however, they run aground in shallows, and are then useful indicators of possible shoals, and submerged dunes. In the moonlight it was terribly difficult to assess the status of these small 'floaters', and I usually gave them as wide a berth as possible.

On an earlier occasion, during November, I had one somewhat alarming experience with a vagrant island of about half a hectare in area. Mohammedu and I had hunted and killed an elephant on a headland north-west of Portofino. We had first approached the elephants in the *Jolly Hippo* and anchored about 1.5km off shore, rowing over towards the herd in the Sportyak. After the hunt, Mohammedu went ashore to a village to arrange for help with the work of butchering the meat and extracting the ivory next day, and remained there overnight, while I remained aboard. During the night a gale blew up from the north-east. The yacht was completely exposed, and anchored in very shallow water. I had a rough night out there, but did not worry unduly until, in the uncertain and shadowy star-light struggling down between the blowing clouds, I realised that a tufted vagrant island was blowing apparently straight towards the *Jolly Hippo* as she lay at anchor. I tried to shift the anchor, but, as a member of the expedition had previously removed the trip line, feeling it to be a nuisance, it was impossible for me to raise it. I decided that if the island actually struck the yacht I should simply transfer to the island and let it carry me ashore. I quickly put on a buoyancy jacket and waited, but the island just scraped by the yacht and fetched up on the shore, near the elephant carcase.

So now Mohammedu and I continued sailing on homewards in the silvery moonlight, through the lagoons, and among the floating islands, during those early-morning hours. We listened to the hippos grunting and guffawing, the chorus of the myriad frogs and crickets, the sigh of the wind in the sails, and the creaming of the water under our bows, and I felt this to be one of the most exquisite experiences of my whole life.

When the dawn broke, the sky blood-red and the high wispy clouds tinged with gold, Mohammedu went for'ard to perform his ritual ablutions, take a bearing on Mecca and say his prayers. He knelt there on his grass kneeling mat, his hands raised in supplication: on this lonely lake, God the Creator and man the creature must commune together.

It was not until the afternoon that we eventually emerged from this labyrinthine belt of floating vegetation

## Chapter 3 - Lagoons, dunes and floating islands

on to open water. The breeze was strong and steady now, the burgee stiffly fluttering at the masthead. As I looked up at it, sitting at the helm, I suddenly saw what seemed to be a compact bunch of very tall, inverted willowy trees in the sky just between the masthead and the foresail, right on our southerly course. There was no horizon now, and the sky and water seemed to be continuous between the prow of the yacht and the trees. But there was no mistaking them: it was the reflected image in the sky of the shade trees of Baga, and we were dead on course.

There were still some five nautical miles to go before reaching the floating islands at the entrance to Baga harbour, but we were now making a good 5 knots, and as we goose-wimged proudly into Baga Harbour under full sail at about 3.30p.m., I saw through the binoculars the foreshore lined with people: it seemed that the whole population of Baga village had come out to see this unusual sight. As we had formerly usually used the motor when entering or exiting the Portofino harbour, and only unfurled the sails when we reached open water beyond the islands, it was their first view of a genuine yacht coming in under full sail into their own harbour.

Catching the strengthening breeze, we turned sharply west, once we were inside the bay, and headed towards our mooring at Portofino, spanking along merrily. We ran in neatly towards our buoy, furling the sails with practised efficiency just at the right moment   With a contented sigh

I thought of my sailing tutors on the Solent, and purred to myself that now at last they could really have approved . . . Mohammedu reached for the buoy with the boathook. He missed it. The *Jolly Hippo* baulked, stopped short, and turned her nose away as if she had encountered a bad smell. I felt dismayed, annoyed and foolish. All the villagers of Portofino seemed to be on the shore watching us.

Mohammedu looked at me accusingly. We unfurled the sails, got into position and tried again. Exactly the same thing happened. Three more times we were thwarted. This *Hippo* was becoming too wild and independent for my liking: she was turning mean! Then, finally, I tried a new tactic. I raced down to the north of the buoy, swung round it to its south, went about, and thus came broadside against it - and Mohammedu caught it.

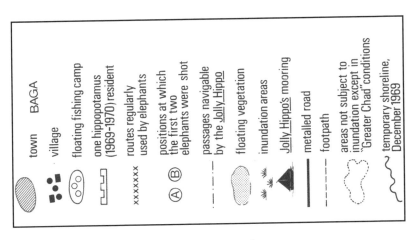

*Map 5: Explanation.*

Chapter 3 - Lagoons, dunes and floating islands

*Map 5. Baga and Portofino : shallow neighbouring harbours bounded on the north by floating islands.*

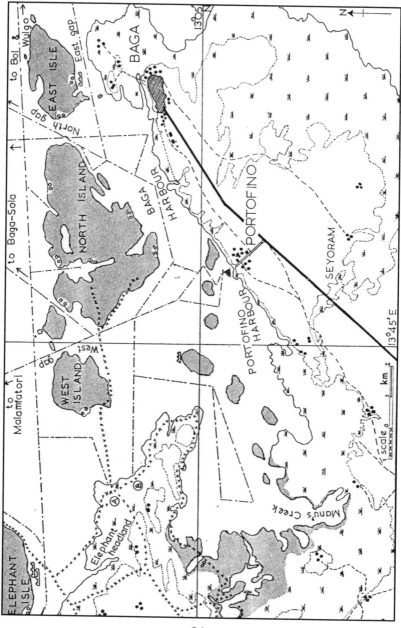

Then the explanation dawned on me and I recalled other similar, though less dramatic, incidents previously when we had missed the buoy two or three times running, and once had even cut the log line with the propeller and lost the log in consequence. On each occasion my fellow crewmen had favoured me with the type of accusing glance that says 'Some helmsmen simply do not know how to steer!' and I had felt annoyed and cheated, and shrugged off their sarcasm with a laugh.

But now it was apparent that this was in fact another sample of those wretched wind-eddies, so characteristic of the Saharan winds, as they snatch up water and sand and form it into wave patterns whereby sand dunes, ripples and channels form. When the site for the mooring buoy had originally been selected at the deepest point in the harbour, seemingly sheltered by the two low grass islands to the north, the focal point of the wind eddies around these islands had inadvertently been chosen, for here three deeper channels also met, formed by the water currents raised by the eddying winds.

When I later took new depth soundings, I found that the three channels formed a deeper hollow in the lake floor at this point of about 4-5 metres in diameter. This also provided the explanation for another episode which had, at the time, puzzled and annoyed us. One morning we arrived at the yacht to find that a fisherman had laid a net in the form of a square completely enclosing the yacht.

## Chapter 3 - Lagoons, dunes and floating islands

This hollow was particularly favoured by fish, and, again inadvertently, we had, by mooring our boat right above it, assumed 'territorial fishing rights' over it. In retrospect I wonder how the Portofino fishermen managed to be so consistently friendly and long-suffering towards us. I can only think that perhaps they imagined that we realised from the beginning why this spot was so special: if so, it was a respect totally undeserved.

So the wonderful months with my yacht on Lake Chad at last had to come to an end. I had sampled the timeless peace of slipping through the silky waters of the languid lagoons and floating islands. I had known the thrill of feeling the Harmattan from the desert fill the sails and send us creaming away across the open reaches. I had experienced the intense satisfaction of setting a successful course by taking a bearing on a floating island, and then homing accurately on a clump of inverted mirage trees suspended in a dazzling sky that had neither beginning nor end. And yet, with all the exhilaration of sailing in this unique setting with its superbly challenging conditions, I am left wondering:

*How is it possible that apparently no one else had ever before brought a genuine sloop-rigged yacht to sail the subtle waters of Lake Chad?*

# Chapter 4

# The puzzle of the puddle

Whether sailing its moody waters, chasing its mirages, or relaxing aboard a floating island, I found on my 1969/70 expedition that I had plenty of opportunity and incentive to ponder the hydrological puzzle presented by Lake Chad. This lake immediately takes many of us by surprise because it is astonishing to imagine:-

*1. **An enormous shallow lake** in the southern edge of the **Sahara desert;** once **the vastest inland water in Africa*** (and sixth largest in the world).

*2. **Its water level rises** in the **dry season*** (September to March) and ***falls** in the **rainy season*** (April to August).

*3.* It has ***no outlet,*** yet the open water remains sweet.

*4.* It is believed to have ***the richest indigenous fish stock in Africa,*** but is a seriously ***fragile ecosystem.***

*5.* It is home to the descendants of a unique and ancient tribe of ***lake dwellers: the Yedina .***

## Chapter 4 - The puzzle of the puddle

In this chapter I shall try to address the first three points (above), leaving the fourth (fish) to chapters 6 and 9, and the last, the Yedina tribe, to chapters 7 and 8. I believe that every serious traveller, geographer or conservationist who reads this book will need and want to grasp how this unique desert lake functions. This chapter was written in 1970 so references to 'nowadays' refer to my knowledge up to that year. Part IV chapters 11 and 12 refer to further events and further research up to the year 2002. Research on the details of the hydrology of the lake have advanced greatly in recent years and are well summarised in the monograph *Hydrologie du lac Tchad* published by ORSTOM / IRD éditions in Paris (Olivry et al, 1996)[6]

### Research organisations and maps.

Over the past sixty years (up to 1970) a great deal of accurate hydrological and related data have been amassed by scientists associated with the French research organisation O.R.S.T.O.M. (Office de la Recherche Scientifique et Technique Outre-Mer), the Federal Fisheries Service of Nigeria, and the Geological Survey of Nigeria. ORSTOM has its impressive laboratories, library and research vessels based at Fort Lamy\* in the Republic of Tchad. The Federal Fisheries Service of Nigeria has a small research station of more recent origin than ORSTOM sited at Malamfatori on the

---

\* Throughout this chapter today's Tchadian capital *N'Djamena* is referred to by its pre-independence name *Fort Lamy*

western shores of Lake Chad near the Yobe River delta, and the Geological Survey of Nigeria operates its Chad Formation Groundwater explorations from Maiduguri. Contributions on the geology of the area have also emanated from the Institute for Agricultural Research Samaru, Ahmadu Bello University, Zaria, Nigeria.

The Lake Chad Basin Commission coordinates major research operations. It is based in Fort Lamy, and represents the interests of the four neighbour countries (the Republics of Tchad, Niger, Nigeria, and Cameroun\*) whose territories impinge and meet on Lake Chad. This body, whose character is politico-diplomatic, is also responsible for making recommendations to the governments of the participatory nations on the economic application of research findings and the conservation of their water resources.†

In spite of the challenge Lake Chad presents there is tolerably good up-to-date *map coverage* of the whole of the Chad Basin, including (on the Nigerian side) complete coverage at 1:50,000 based on ground and aerial surveys of 1951-63, and (on the French-speaking side) 1:200,000. In addition there are various older maps of varying degrees of accuracy, but nonetheless of great interest as providing important historical data about Chadian hydrology.

---

\* Now in 2002 it also includes the Central African Republic.
† The latter function will be further mentioned in Chapter 12

Chapter 4 - The puzzle of the puddle

## 1. An enormous shallow lake in the southern edge of the Sahara desert.

Part of the fascination and challenge of the enigma of Lake Chad is its geographical and geological background. Then there is its climate: generally very hot and sunny, except when either the north-easterly *Harmattan* wind obscures and cools everything under a white diatomaceous haze, or else localised storms drench localised areas.

Seen from the air, or from space, the lake can be recognised as a roughly circular wetland with extensive open water expanding into two 'basins', and lying roughly at the centre of an expanse of dune-lands or *ergs*. *(See Map 3 page 4)\**. The climate is extremely hot with air temperatures usually in the range 25 to 40° C, with occasional falls at night to a minimum in January of 15°C *(See Diagram 2)*. Surface evaporation of the lake water around 1969/70 was estimated as shown in *Diagram 3*. The southern basin, during our expedition was about 180km by 75km in size, while the northern basin was about 140km by 95km. This expanse provided us with some really long voyages across open water in the *Jolly Hippo (Plate 3.1)*.

The lake's vast area, shallow in depth, is erratically fed by one very large *(Chari)* and three small, intermittently flowing, rivers *(Yobe, El Beïd and Yedseram)*.

---

\* See Chapters 11 and 12 for details of more recent changes.

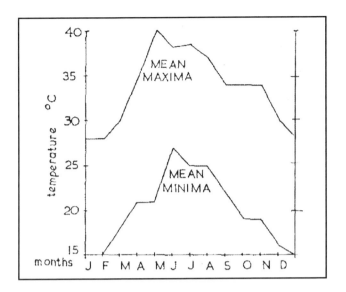

*Diagram 2. Mean monthly air temperatures (maxima and minima) of the northern basin of Lake Chad.*

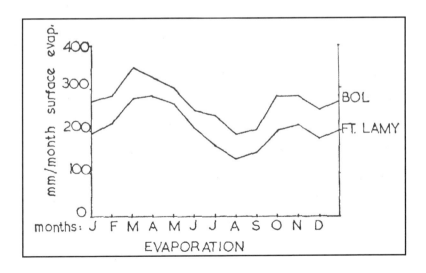

*Diagram 3. Mean monthly surface evaporation on Lake Chad at Bol and the Chari River.*

## Chapter 4 - The puzzle of the puddle

The large Chari River enters the lake through a delta at the southeast of the southern basin, the smaller Yobe through a delta at the northwest of the northern basin, and the Yedseram and El Beïd at and near the southern limit of the southern basin.

Because the lake is relatively so shallow and the shores so gradual, any significant rise in level causes extensive increases in area of the whole lake carrying its dissolved salts and suspended silt outwards with the overflow. This has significant consequences to be discussed later. Such inundations have been historically called lake 'transgressions'.

An almost straight line arises on the extreme eastern corner of the southern basin: this is the *Bahr\* el Ghazal*. It is a water course extending out across the desert from the lake, usually completely dry, but sometimes containing pans of water or even becoming inundated by lake overflow when the lake level becomes high, as in 1870, 1873, 1874, 1900, but not again until 1956, following the heavy flooding of the Chari river in 1955, and again in 1957.

It may, in the distant past, have had a positive flow of water *towards* Lake Chad. In fact local tradition suggests that, during Noah's Flood, waters from the Nile Basin flowed down this depression into Chad, whence the derivation of the word 'Bornu', or *'Bahr el Nuhu'*, meaning the 'flood or river of Noah'.

---

\* Bahr: a water course.

Nowadays, however, when inundation occurs, it does so by a *backflow* from the lake up the Bahr el Ghazal towards the north-east.

Water persisted in the depression throughout 1958, but then declined. A definite direct relationship has been established between the limit of inundation of the Bahr el Ghazal and the lake level in any given year as measured at Bol on the lake, and at Tagaga in the Bahr el Ghazal.

The lake bed is not flat, for it lies on an ancient bed of fossilised sand dunes, many of which surface as sandy 'desert islands' when the lake level falls, while others remain hidden and form 'anchorages' for floating vegetation. Concentric underwater sand bars spread away around both the Chari and Yobe River deltas eventually backing up towards the 'Great Barrier' which is a wide bar running south-south-west/north-north-east across the lake between the Baga peninsula (in Nigeria) and Baga Sola (in Tchad).

When the rivers are in spate at the end of the rainy season, a swirl of water in each basin is set up and this circulates chemicals and silt. The chemical situation in the lake waters is measured by conductivity measurements and is important in relation to both the micro- and macro-flora. Oxygen content is also particularly important to the micro- and macro-fauna (especially fish), and its gradients from surface to lake bottom in different parts of the lake are also measured, as are sunlight (insolation) and

## Chapter 4 - The puzzle of the puddle

moonlight, surface temperatures, and relative humidity which affect plankton movements and growth, and invertebrate life-cycles.

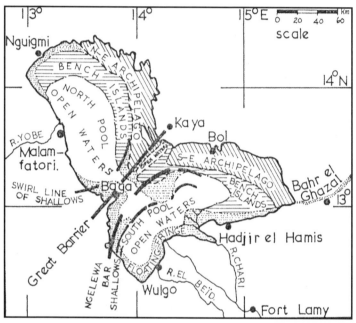

Map 6. The topography of Lake Chad

During our time at the lake, we cruised with great interest through the lagoons and channels overlying both the Great Barrier and the crescentic bars derived from the Chari and Yobe deltas, as well as across the open waters between them. Larger vessels such as the Mission launch *Albishir* could not at that time penetrate many of these lagoons because they were too shallow, but the *Jolly Hippo* could negotiate most of those that we attempted. (Plate 3.1).

On a typical cruise in the *Jolly Hippo,* following the northern shore of the Baga Peninsula on a course of 80°, we would first cross Baga bay, a distance of about two nautical miles of shallow water then rarely exceeding 1m depth, with the bottom covered by a thick layer of humic, muddy 'ooze'. Leaving the harbour by the northern exit *(Map 5 p.84)* between floating papyrus islands, the channel would deepen to about 2 metres before we encountered open water, and then we would cross some 8km of open water with a firm sandy bottom varying between 2m and 3m in depth, with scattered low submerged bench islands. Portions of the trace of our graphic echo sounder of this course are shown in *Diagram 4, pages 96-98..*

Heading across these open waters we would soon see the low line ahead marking the beginning of the floating vegetation. As we approached, some circular, anchored floating islands would first become distinguishable, and then an apparently continuous line of dark green floating vegetation anchored to the muddy substrate by its roots and initially appearing to be absolutely impenetrable.

Approaching closer, however, various channels would become apparent, and, selecting those nearest to the intended course, plotting any diversions carefully, we pressed on into the maze of lagoons.

## Diagram 4. Sample traces recorded by the echo sounder of the Jolly Hippo (pages following).

(a). Inside Baga harbour following a course of 80°. Very shallow water, the bottom consisting of thick humic ooze.

(b). 1 and 2. Open water, north of Baga harbour with a firm sandy bottom and long, low bench islands or shallow bars (see map 6).

(c). 1 and 2. Shallow lagoons among floating rafts of Phragmites reed-grass and papyrus, some 10 nautical miles from Baga, on a course of 50º

3. Low bench island or bar in the N'gelewa Bar area about 18 nautical miles from Baga. Neighbouring bench islands and bars in this area carry rafts of anchored, floating vegetation, and the bottom is generally covered with a soft layer of muddy humic ooze. Depths here ranged to 4.5 m in January 1970

(d). 1 and 2. Bench islands in the eastern side, on approximately the same latitude as Malamfatori, on a course of 70°. In some cases the bench islands of this area arise somewhat abruptly from the lake floor and, where they lie within 1.5m of the lake surface, carry a covering of aquatic water weed. Towards the eastern end of this run, exposed dune islands were also encountered. Below the floor of the lake in these traces, secondary echoes suggest an underlying sandy bottom.

(e). 1 to 4. Eastern part of the run mentioned above (d), showing depths of as much as 6m, well-defined bench islands, generally with a steeper slope on the eastern face (left here) than the western face

(f) 1 and 2. Scattered bench islands in the open waters of the north pool on a course of 3°, about 25 nautical miles north of Baga. Note the firm sandy bottom.

(g). 1 to 3. Open waters of south pool on a course of 20°, some ten nautical miles from Fodio Island, showing fossil sand dunes underlying clay deposits forming a level bottom surface some 2 to 3m deep.

## Diagram. 4. (a) to (c).

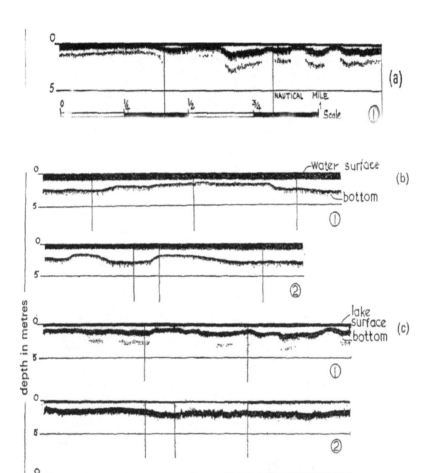

Chapter 4 - The puzzle of the puddle

## Diagram 4. (d) to (g).

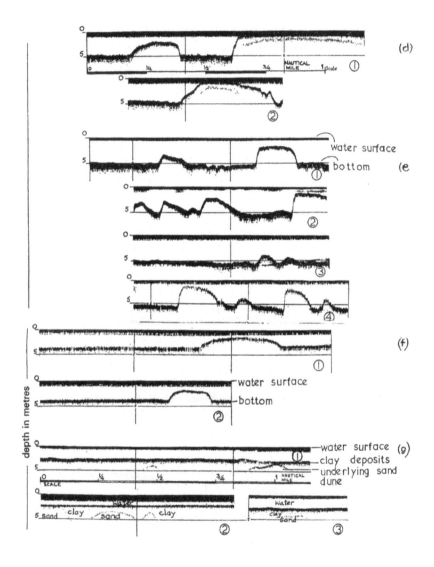

## PLATE 5
## LAGOONS, DUNES AND FLOATING ISLANDS (C)

1. *(left)*. Yedina fishermen in their regular papyrus workboat or *kadai*.
2. *(below)*. The author with Yedina fisherman: navigation was always difficult in 1969/70 because the GPS system was not then available for common use. The ever-changing lake level and vagrant floating islands made recognition and confirmation of co-ordinates uncertain. Using Ordinance Survey maps, guesswork and the Yedina's knowledge, it was possible to relate estimated position with actual dune islands.

3. *(above left)*. Dawn on Lake Chad, with an imported dugout canoe and fish net markers.

4. *(above)*. Setting out to the fishing grounds, in the early dawn light, when water and sky merge.

5. *(above)*. Sunset on Lake Chad when you can sense its strong mystique

**PLATE 6**

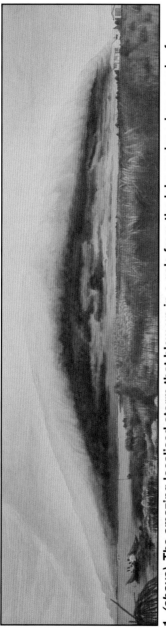

1. *(above)*. The amazing localised storm that blew up in a gale from the lake under a huge arch of advancing cloud in August 1969. It rolled on over our heads pouring down torrents of rain while, above the dome, the sun reflected brightly upwards. *(Painted by Mary Thomas from a panoramic sequence of the author's photos)*.

2. *(above left)*. Hurrying ashore as a storm approaches. 3. *(above right)*. Crossing a lagoon in calm weather.

# PLATE 7
# THE FLORA & FAUNA (A)

1 & 2. *(above).* Papyrus *(Cyperus papyrus)* and mauve convolvulus *(Ipomoea rubens)* are both common as the floating vegetation of vagrant islands and as shallow-rooted fringe vegetation of the mainland shore and of southern dune islands.

3. *(above left).* A thick stand of Ambach *(Aeschynomene elaphroxylon)* in the background, and water lilies *(Nymphaea sp)* covering the shallow waters of the lagoon. 4. *(above right).* Doum palm *(Hyphaene thebaica)*

5. *(above left).* White herons in the shallows with floating convolvulus, note ambach beyond. 6. *(above right).* Pelicans above the lake (1974).

# PLATE 8
# THE FLORA & FAUNA (B)

1. *(above left).* Ground hornbill. 2. *(above centre).* A speckle-throat otter hand-reared by the author: they were fairly common on the lake. 3. *(above right),* A pet stork kept by a Yedina fisherman.

4. *(left).* Pet ostrich in Yedina compound.

5. *(right).* Abdim's Stork

6. *(above).* Sitatunga bull.

7. *(above).* Female sitatunga calf hand-reared by the author.

*Lake Chad versus The Sahara Desert*

# PLATE 9
# THE FLORA & FAUNA (C)

1. *(above left).* 'Twinkle' a red- fronted gazelle (female) hand reared by the author in the same cage as the leopard cub 'Pumpkin' at Lake Chad (1969).
2. *(above right).* Wild oribi near the shore of Lake Chad (1969).

3. *(above left).* 'Ginger' the baby patas monkey who became the constant companion and beloved friend of orphan leopard 'Pumpkin' ( later re-named 'Sultan'). 4. *(above right).* 'Pumpkin' plays with cage companion 'Twinkle'.

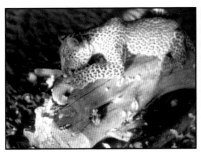 

5. *(above left).* The leopard cub's first pigeon. He seems to say: 'Whatever should I do with those feathers? But I know what to do with the flesh!'.
6. *(above right).* Tearing a *kadai* to pieces was fun.

# PLATE 10
# THE FLORA AND FAUNA (D)

1. *(right)*. At one year old old it seemed as though 'Ginger' thought he was a leopard, and 'Pumpkin' thought he was a patas monkey: they were completely devoted to one another.

2. *(below left)*. An amazing relationship.

3. *(left)*. Tiger fish.

4. *(right)*. Nile perch, or *Giwan ruwa* (water elephant), the largest fish we saw caught while at the lake, although larger ones are reported

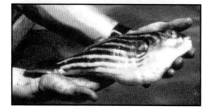

5. Puffer fish *(above left)* deflated, and 6. *(above right.)* inflated.

## PLATE 11
## FISHING (A)

1. *(right).* Bringing in the nets for repair.

2. *(below).* Fish net repairs.

3. *(above).* The fishing fleet at Portofino.

4 and 5. *(above).* Preparing fish for drying: gutting and chopping into chunks and then spreading them on papyrus or grass stem platforms for sun-drying. Sometimes a smokey fire is also lit underneath.

## PLATE 12
## FISHING (B)

1. *(right).* Small whole fish are hung on lines in the sun. All the drying fish attract hordes of flies, which impair the quality. This detail does not appear to reduce their value or to prevent their export by road all over Nigeria.

2. *(above left).* This fisherman favours riding astride an ambach plank with a gourd at each end, and rows it along with his hands.

3. *(above).* A heavily laden dugout canoe with sacks of dried fish, assisted by a sail, is poled through shallow water. 1974.

4. *(above left).* Fishermen go out to the fishing grounds early. In the 1974 drought it was easier to ride these gourds rather than to struggle on foot through the muddy shallows.

5. *(above).* Riding on a gourd-boat, a fisherman inspects his lines.

Frequently such channels would be laced with nylon gillnets, and nylon lines laden with fish hooks, the fishermen, dotted around in papyrus or dugout canoes, tending them. Here and there a Yedina floating camp might be visible, waving up and down gently on its floating platform of papyrus and *Phragmites* rhizomes\*.

Cruising along through these channels and lagoons, we sometimes encountered extremely shallow parts lined with a particularly glue-like layer of black saturated mud. These provided tough going for the engine and propeller which left a long black muddy wake trailing far away behind the vessel. On one trip in this area, we found ourselves in a particularly shallow and gluey channel of this kind and were somewhat startled to see a black 'wake' set out from one 'bank', apparently unattached to any vessel, and heading straight across our bow. With a 'hippo ahoy!' we tried to give chase, but although we managed to keep station alongside, we could not, with the resistance of the mud, actually overtake the submerged creature.

Suddenly the broad head and snout, completely covered with absolutely black mud, of a huge bull hippo surfaced, stopped and gave us a very long concentrated stare. Then he ambled off, head half-turned to keep an eye on us over his shoulder, towards the other 'bank', leaving his persistent and continuous muddy wake now stretching

---

\* These camps are more fully described in Ch.8

*Chapter 4 - The puzzle of the puddle*

*Diagram 5. ...my eyes met those of a hippo glaring at me.*

from side to side across the channel. As there were no fish nets or fishermen anywhere in the vicinity of this rather tightly-packed series of shallow channels and cul-de-sacs, and as there were a number of distinct muddy channels leading in and out of the floating vegetation, we supposed this to be a favourite, and generally undisturbed, residential area for hippos. Graphic depth traces of this lagoon area are shown in *Diagram 4(c)*

On the approach to the Tchadian shore along the Grand Barrier, papyrus, *Phragmites* and reed rafts gave way to very low, submerged, and partially submerged bench and dune islands, the latter clad with low grasses and the inter-dune channels filled with water weed. It was here, as well as in the north-eastern archipelago, that in February we encountered small flocks of migratory waterfowl, of which the most abundant species was whistling teal.

Nearer to the Tchadian shore itself are long narrow dune islands orientated north-north-west/ south-south-east bearing a sparse cover of doum palms and grass, and fringed with floating vegetation or low sedges and convolvulus.

When the Great Barrier became an exposed series of sand ridges in 1904 to 1908, it was already somewhat higher than it is now. But then, when it was re-submerged, it was eroded by water action to assume its present level resembling a series of long low bench islands, except for its two ends at Baga and Kaya. The Ngelewa Bar (or Ridge), on the Chari side (on which incidentally the intersection of the international boundaries of the Republics of Tchad, Cameroun and Nigeria meet), is lower and was apparently not fully exposed in 1904-1908. It is traceable with interruptions, from near Mongonu on the west side of the south basin to Gambaru on the El Beïd river on the east - in fact it was the starting point of my earliest experiences of Lake Chad in 1955. It stands about 3m above the surrounding inundation plains *(firki)* and is evidence of the former limit of Mega-Chad, the former '320m' lake.

So eroded are some parts of the northern end of the Great Barrier today that several useful navigation channels can be located running from Bol to Tataverom and Magi, one of these passing alongside of the south-western shore of Samiya Island where the *Jolly Hippo* first broke down and was temporarily abandoned.

These are at present deep enough for navigation by motor barges, as well as by the Mission and Research launches, plying between Bol, Haikalou, Tataverom and Magi.

The dune island archipelagos of the north-eastern sides of each of the two pools of the lake consist of literally innumerable low elongated fossilised dunes aligned north-north-west/south-south-east. These originally formed during periods of extreme aridity in Quaternary times, as a result of wind action on sand forming a vast dune field. Prevailing winds and cross-winds today appear to have the same direction as then and tend to maintain not only the same dune shape even where they are now exposed to erosion, but also to set up local surface water currents which maintain the inter-dune channels, so that many of them remain relatively clear of silt and weeds, especially in the more exposed areas towards the open water of the lake.

*Diagram 4(d)* shows a depth trace taken on a voyage in the north basin in December 1970 on a bearing of 70° where exposed dune islands alternated with submerged bench islands. Where bench islands lie within about 1-1½m of the surface of the water, they are often covered with a floating raft of well-rooted *Phragmites*, papyrus or reeds. Apparently this cover may become detached if the depth of water exceeds 1m. In the past ambach trees also grew around many of the dune islands but died out earlier this century, leaving only scattered relict stands today *(see Chapter 11)*. These were used a lot in the past by the

Yedina people to make a kind of surf-riding plank which they propelled with hands and feet while lying full-length along it on the water. These can still occasionally be seen around Bol but the 'surf-boards' have given way almost completely nowadays to plank-built and dugout canoes, and the occasional gourd-rafts described in Chapter 1.

The open waters form two broad expanses, one in the western part of the north basin and one in the southern part of the south basin, the channel within the Great Barrier forming the only other significant stretch. The extent of the open waters naturally depends upon the prevailing surface level of the lake, for, the deeper the lake is, the fewer are the exposed dune islands bordering them, and, in addition, the floating vegetation declines and becomes progressively reduced as the lake surface rises, increasing again as it falls.

In *Diagram 4 (e) and (f)* depth traces taken in the open waters of the north basin on a bearing of $3\frac{1}{2}°$ and in the south basin on a bearing of $20°$ are shown. In the open waters of the south basin we several times encountered nylon gill nets exceeding one nautical mile in length, suspended from stout wooden poles along the underwater summits of ridges running in the same alignment as the Ngelewa Bar. Although we did not check the length and entire course of any of these ridges, they seemed to lie in a north-north-east/ south-south-west direction, that is parallel to the Ngelewa Bar and presumably represented long-distance offshore water action from the Chari delta.

## Chapter 4 - The puzzle of the puddle

Our first impression of these nets was that they had been placed at random 'in the middle of nowhere', but as time went on we soon discovered that the Yedina fishermen showed tremendous fishing acumen over the siting of their nets, an acumen apparently not shared by the Government Fisheries' personnel whose net siting was not infrequently the butt of Yedina contempt and mockery. In practice, the Fisheries' nets were said, by the fisheries officers, to be sited in 'experimental' positions.

The southern shores of the south pool and the shores of Baga Bay (i.e. the northern shores of Baga Peninsula) are heavily fringed with floating vegetation, the papyrus rafts being particularly luxuriant here, but still of the same general height and type as those capping bench islands. Here they are rooted in off-shore shallows and spits and form an ever-changing maze of lagoons, inlets and floating islands. Here, too, the finest *kadais* (papyrus canoes) are made by the Yedina fishermen. Recent Hausa and Kanuri settlements also occur along the shores behind the papyrus fringe of floating vegetation. These are either populated by immigrant Hausa people from around Sokoto, in north-western Nigeria, or Kanuris who have adopted fishing as their substantive occupation. These settlers do not make *kadais* and rarely use them, preferring large dugout canoes imported from the Benue and Niger Rivers by traders and sold at the lake shore for between N £20 and N £30, according to size.*

---

* Nigerian pounds later replaced by the less valuable Naira

As mentioned earlier, the rafts of floating vegetation capping bench islands, shore-line shallows and sand spits sometimes fragment either owing to the dry season lake rise, when roots are fractured or simply uprooted from their sub-aquatic substrate, or else due to high winds and storms in the wet season, which literally tear the floating rafts apart. These fragmented floating islands are locally known in the Yedina language as *kirta,* and usually assume a circular shape after detachment. They blow about with the wind until eventually they become wedged between two anchored rafts or islands, or become stranded on a bench island, shallows or sand spit. Then they take root.

### Bottom sediments.

The actual substrate comprising the lake floor is also of the utmost importance as a biotic environment. The bottom deposits which clothe the actual lake floor are essentially clayey or sandy. These are;

*(i) Grey to black clayey mud,* saturated with water and rich in organic matter consisting of vegetable deposits of about 1mm in size. This deposit consists of about $1/3$ clay, $1/3$ silt, and $1/3$ sand. This was absolutely horrible to stand in when we had to climb out of the yacht to clear the propellor of fish nets.

*(ii) Grey-blue, polyhedral clay.* This is a granular clay consisting of polyhedral angular grains ranging in size from 0.5cm to several centimetres across. It consists of one-fifth sand, two-fifths silt, and two-fifths clay.

*(iii) Hard, blue clay* with fragments up to 30mm across, rounded edges which are very difficult to break and in texture resemble coral or shell sand.

*(iv) Fine sand, aeolian\** in origin with grains up to 0.15mm diameter, very little organic matter, and frequently with a scattering of dead mollusc shells.

*(v) Pseudo-sand* : this is really a type of granular clay which looks like coffee grounds, and may form a layer up to 40cm thick.

*(vi) Peat* occurs around the shores of many dune islands in the archipelagos, and may be several centimetres thick.

The echo sounder used by the workers of ORSTOM on their research vessel frequently shows a double echo. These echoes represent two layers of distinct bottom sediments. Generally, they reveal a soft layer of clay, silt or peat overlying the hard sandy layer of the fossil dune field which partly underlies the eastern and northern parts of the lake. Our own echo sounder was not generally able to pick up the double echo but it may be seen here and there in *Diagram 4(g)*.

---

\* Wind-borne

The fauna characteristic of these different types of bottom sediment are described more fully in Chapter 6. Although the lake bottom has by no means been fully explored up to the present, one thing is certain: the clay and sandy clay bed of the lake forms an effective impervious layer through which the waters of the lake cannot possibly percolate downward to re-charge pressure aquifers and there is absolutely no surface or sub-surface escape channel through which they can flow. Some percolation into and along unconfined aquifers (water table) is possible but very limited, and is seen in some inundation areas.

Whether there are any *upward* seepage points of pressure water from deep underground aquifers existing within the lake bed is as yet unknown, but this is a possibility. Indeed, there is said to be a bottom 'pit' on the western side of the north pool where there is clear water, and where luminous gases may be seen at night. Nevertheless, it seems unlikely that, if any do exist, they affect the lake level significantly.

### The location and history of Lake Chad.

*Maps 2, (page xxiv) and 3, (page.4),* with the map inside the *Endcovers,* help to locate Lake Chad. We see three outlines in *Map 3* showing rather simplistically the supposed history of the lake as follows:

### c 55,000 years ago: Palaeo-Chad.

Around this time apparently the Chari River was very large and active, and formed a great delta covering 40,000km$^2$ in association with a vast lake or freshwater inland sea of which the surface stood at 380-400m above mean sea level (AMSL) and which occupied about 1.95 million square kilometres.

This was *Palaeo-Chad*, occupying a basin which was being structurally modified by the deposition of sediments but has been geologically relatively stable since Pliocene ('Early Recent') times, and possibly even earlier, in Tertiary ('Pre-Recent') times and now known to geologists as the *'Chad Formation' (Map 3, page 4)*. Pias[7] hazarded a guess for the age of this of about 55,000 years.

### c. 22,000 to 12,000 years ago. Early dune formation and lake regression :

During one or more extremely arid periods following, the vast Palaeo-Chad receded and moving *sief* (sword) dune systems formed between latitudes 11° and 20° N under the influence of seasonally-prevailing winds from the north-east and north-west. These are still the general directions of the prevailing winds. The north-easterly wind, known to us as the *Harmattan*, is characterised by its heavy burden of fine diatomite dust which creates a haze in the atmosphere. So any changes undergone by the basin or the lake may be supposed to reflect primarily the climatic changes both of the Sahara itself and of the country to its south.

Dune movement was prevented from sliding further west by a north-south line of quartzite hills near Katsina in Nigeria and the Goudam hills in Mali. Towards the western side of the basin, in the lee of the hills, lakes formed in hollows leaving strips of lateritic ironstone exposed, which accumulated light sandy soils between them, and which now form useful arable areas. But right across the centre of the wide Chad basin two great dune systems formed: the 'great erg of Hausaland' and the 'great erg of Kanem'.

### *c.12,000 to 7000 years ago. Mega-Chad: the Pluvial transgression.*

Humid conditions apparently followed the arid era, and the lake again enlarged, this time with its surface just under 320m above mean sea level and occupying the centre of the overall Chad Basin. The peak of this period seems to have been about 10,000 years ago. There is visible evidence for this era, provided by Migeod's strandline. *(See Map 7, page 113, for strandlines or ridges).*

During this period the Chari River formed a new delta north of present-day Fort Lamy. while south-flowing tributaries from the Niger Republic, with east flowing rivers from Katsina and Kano, would have joined the Hadejia-Yobe system, transferring the earlier Yobe delta to the position near its present delta, where it is still visible as a raised area of coarse sand. Carbon-dating of diatomaceous material deposited in oasis depressions between 300 and 400m above mean sea level shows that the earliest were deposited in the Palaeo-Chad era 21,350±350 years ago, and those in more recent layers 9,000 to 7,000 years ago.

### c. 7000 to 5000 years ago. The lagoonal period: lake regression to just below 287m above mean sea level, or the Ngelewa level.

Authorities differ in their interpretation of this period, Pullan[8] suggesting that Mega-Chad persisted around its 320m level to as recently as 5,000 years ago. Others, however, suggest that this was a period of regression, or perhaps wide oscillation, of the level of lake waters, as there is evidence that there were numerous lagoons and pans north and east of present-day Lake Chad, and that the area was well vegetated.

Neolithic artefacts occur above the 300m level suggesting that during this period the lake may have oscillated about a mean level of just under 287m above mean sea level - i.e. the level suggested by the Ngelewa Ridge. During this period the Chari River had formed a bar inside the lake, beyond its delta, enclosing a large lagoon behind it. This may have been heavily vegetated as it accumulated humic and silty clays in just the same way that the Great Barrier across Lake Chad does.

***c.5,000 to 2,500 years ago. Lake Chad: regression to present (1970) lake level at 280m above mean sea level.***

It was probably during this period that the lake assumed its current average level with periodic oscillations involving rises and falls of a few metres from time to time. Barchan (crescentic) dunes, which are mobile even today, began to form at this time in the Bodele Depression. The Bahr el Ghazal, now normally a dry channel running north-east from Lake Chad, probably flowed and dried up alternately during this period, with swampy intervening periods. Seen from the air, it is immediately obvious that today's lake lies at the heart of a complex of wide fields of sand dunes, and also that there are some peculiar ridges which can be identified as the strandlines of an ancient inland lake or sea *(Map 7, p113; map 8, p116)*.

Dunes may cease to be mobile and become compacted and 'fossilised' as conditions change and this has happened to many of the dunes in the Chad basin, notably those actually underlying Lake Chad itself. This suggests forcibly that at some time in the past during its evolution from *Palaeo-Chad,* an actual lake simply did not exist over one or more periods. Perhaps at some time, during its 'transgressions' and 'regressions' ('oscillations') it was so desiccated that the dunes invaded most of its bed, a theory proposed by Falconer[9] as early as 1911.

He associated origins of the wide clay plains to the west

and south of the lake with a period of higher rainfall. Unfortunately he failed to recognise that the significance of the prominent sand ridge passing through Maiduguri and Bama was associated with a former southerly limit of the lake and was in fact an ancient strandline. This is now known as *'the 320m ridge' and represents the limit of 'the 320m lake' or 'Mega-Chad'*.

**The sand ridges**

Then in 1924 Migeod[10] did notice the prominent discontinuous sand ridge about 12m high which surrounds Lake Chad at a variable distance from its shore line and interpreted its existence and character, together with the ancient deltas and clay plains in terms of former 'transgressions' and 'regressions' of the lake.

The 12m high strandline, with its summit at 320m above sea level, runs across the plains of Bornu from north-west of Magumeri, through Maiduguri and Bama (Map 7) and south-east to Yagoua on the Logone River opposite Bongor. A further sandy ridge runs east from the Massenya area, east of Lake Fitri and then north across the Goz Kerké, a little to the east of Koro Toro, near Faya. It continues as the well-defined Taimanga Ridge. Further west the former lake shore is bordered by alluvial spreads laid down by defunct streams that used to arise in Tibesti.

---

\* More recently Known as 'Conventional Lake Chad'

A field of *barchan* dunes obscures the position of the shoreline beyond this until the Hero Maro Ridge is reached which runs south across Ténéré to the high dunes of Kanem. Near Nguigmi, a bare sand ridge crossing the lower end of the Dillia Wadi clearly marks the former lake shore. This ridge provides evidence of the limits of the former '320m' lake, or Mega-Chad.

An inner and smaller ridge, the Ngelewa, (described earlier) is seen well at Gambaru where it stands about 3m above the surrounding plains, except where blowing sand has, in some parts, drifted against it. Remnants of another

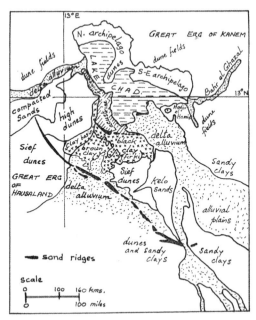

*Map 7. Geomorphological units relating directly to Lake Chad and its feeder rivers, showing strandline ridges.(after Pullan)[11].*

ridge between Bir Kerala, Tourba and Al Greg to the east of the present mouth of the Chari River, probably coincide with the Ngelewa Ridge and represent the limits of a more recent, reduced lake with its surface just under 287m above mean sea level.

**Hadjir el Hamis.**

No rocks or stones occur in or near Lake Chad, with the single exception of the igneous extrusions of the basement complex (which elsewhere underlies the Chad Formation) known as Hadjir el Hamis (popularly called 'elephant rocks') on the northern bank of the Chari River, near its delta. *(see Plate 27.2)*. This situation has obvious consequences for people living on the lake itself, for example affecting the way they sink their fishnets, and cook on their kadais.

*2. The lake waters <u>rise</u> in the <u>dry season</u> (September to March )and may inundate the lake shores, and <u>fall</u> in the <u>rainy season</u> (April to August): these are the '<u>oscillations</u>' of the water level of Lake Chad.*

If you are planning to visit Lake Chad, and mention it to people who are aware of its location and existence, they usually say *'Oh! That's the lake in the Sahara desert, that goes up and down in the wrong seasons and has floating islands inhabited by fishermen.'*

It is a good summary, for the lake's water behaviour is not only unusual and distinctive, but also determines the life and survival of the people who depend on it. Its *'transgressions'* and *'regressions'* have already been mentioned, and reflect the fact that it lies in a very wide, shallow catchment basin and is itself relatively very shallow, with a very gradual shoreline over which it can readily overflow. But the most significant and obvious feature is the 'oscillation' of its actual water level, generally identified as the level of the water surface 'Above Mean Sea Level' (AMSL). This form of measurement has been available to scientists for a very long time, the earliest relating to Lake Chad having been made at Bol in about 1872. This also provides the sound basis for comparisons with the historic levels of Palaeo-Chad and Mega-Chad derived from their identifiable hard strandlines or ridges. *Diagram 6, page 119,* shows graphically a reconstruction of annual maximum and minimum surface level oscillations from 1872 to 1970, the latter being the time when my expedition on the lake was in progress.

This graph gives a very revealing picture, demonstrating why, in 1955 ('Great Chad') I experienced that amazing flooding around Wulgo and across the plains from Dikwa to Ngala ('transgression')and why we found yachting on the lake 1969/70 ('Medium Chad') such a magical experience.

## Chapter 4 - The puzzle of the puddle

It was the famous French administrator General Tilho[12] who first attempted to gather data and analyse the cyclic variations of the waters of Lake Chad. He described its oscillations at three general levels, namely *'Little Chad'*, *'Medium Chad'*, and *'Great Chad' (Map 9 and Diagram 6, page 121).*

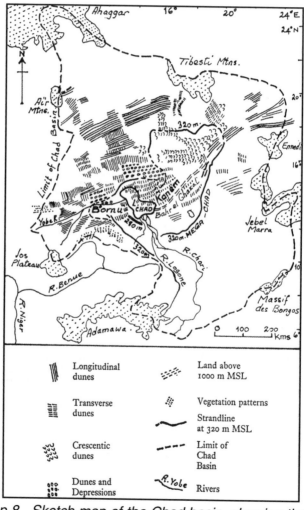

Map 8. *Sketch map of the Chad basin, showing the main dune fields (after Grove and Warren)*[13].

*(i) Little Chad (1905-8 and 1916)*

The lake resembled a vast swamp where small areas of open water persisted, encircled by forests of ambach trees *(Herminiera elaphroxylon)\**. These trees can only grow in very humid conditions and occur among dense reeds where the water does not exceed a maximum depth of 2m. On the eastern side *bahrs* formed into stagnant pools of brackish water. Many of the islands merged with *terra firma* and the fishing and pastoral populations migrated towards the lake centre in search of fresh water and green pastures.

The desiccation of the north basin was so pronounced that fish died in large numbers and lay rotting on the sand or decomposing in foul stagnant pools. When the Great Barrier becomes exposed in the 'Little Chad' phases, it is found that the lake level in the northern basin is lower than in the southern basin, so that the desiccation of the northern basin is more rapid than the southern basin[†].

*(ii) Medium Chad (1909-13; 1917-20)*

Water again occupied the north basin, but the bahrs on the eastern side of the lake were still generally dry. The western and southern shores were somewhat ill-defined, as they were fringed with a very extensive cover of papyrus, reeds and grasses. The forests of ambach died and progressively disappeared from the centre of the lake

---

\* Now renamed *Aeschynomene elaphroxylon*.
† See the effects of this in Plate 26.

towards the shores, leaving only a few stands on the eastern side in the archipelagos.

*(iii) Great Chad (19th century and 1957; 1962-5).*

The bahrs on the eastern side were full and the southern and western shores inundated. The archipelagos had some forests of ambach but the floating vegetation fields and rafts were comparatively small and the shore fringes narrow. Bench islands were deeply submerged, and peninsulas were islands. Navigation across the open waters was easy and speedy. Through 1953 to to 1968 the lake remained 'Medium' except for the notable rises in 1957 and 1962-5 when it again assumed 'Great Chad' levels.

Measurements of depth and calculation of volume and surface area were systematically begun in 1962 by the scientific 'Mission Logone-Chad', and have since then been carried on by ORSTOM. Tilho, however, was the first to install a fixed depth scale at Bol in 1908, and a continuous record has been maintained at Bol since then. The zero mark on the depth scale at Bol is now (1970) the standard of lake level measurement for the whole lake and a surface level of 282.00m above mean sea level - i.e. the 1956 level - is accepted as the base reference to which all changes in surface level may be related.

*Diagram 6. Reconstruction of the surface level oscillations of Lake Chad, showing annual maxima and minima over the past 100 years (derived from actual measurements and correlation with oscillations of the Nile and Chari rivers).*

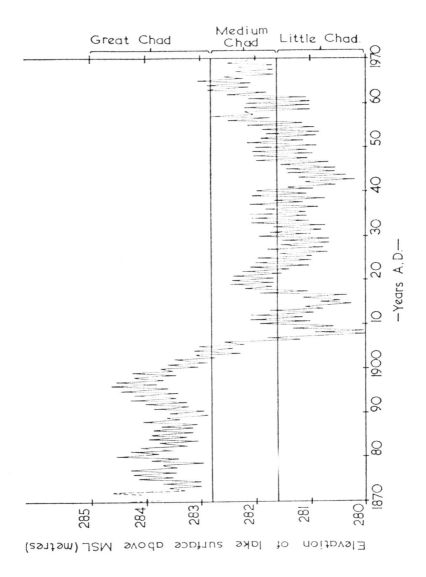

On the basis of these measurements, the average 'Great Chad' lake depth is found to be 4-7m in the north basin, with maxima up to 12m in certain channels in the archipelago. In the south basin depths diminish as the Chari River delta is approached, being about 3-4m in the estuary itself and up to 12m further out.

One particularly deep pit in the lake bed in a channel near Bol is 15m deep. One of the most striking peculiarities of the hydrology of Lake Chad is the lake's extreme shallowness in proportion to the area it covers, so that it is best regarded as a wide puddle rather than as a true lake.

Across the Chad basin lies an extensive and complex **ground water system.** Groundwater comprises *(1) the water table, i.e.* unconfined water, recharged mainly by rain, and river and lake flooding, *(2) pressure (confined or artesian) water*[14] underlying one or more impervious underground layers of clay, or rock, which is probably never recharged in the vicinity of the lake, or, if at all, in remote locations via semi-permeable faults. Indeed, if there is any interchange between the pressure aquifers and the water of the lake, it would only be by *upward* seepages under pressure into the lake from below. The former can be raised for irrigation, domestic, or industrial uses by means of dug wells with buckets or pumps.

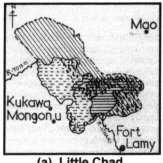
(a) Little Chad

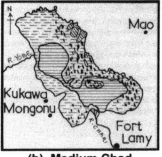
(b) Medium Chad

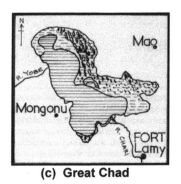
(c) Great Chad

Map 9. 'Great', 'medium' and 'Little' Chad (After Tilho et al).
(a) Little Chad 1908: northern basin dry.
(b) Medium Chad about 1920.
(c) Great Chad 1957: extensive open water in both basins.

The latter, artesian or 'fossil' water is raised by means of borehole pumps, or, if the pressure is adequate, it will rise to the surface within borehole wells under its own pressure.

In the case of the borehole at 'Minetti' *(See Plate 23.1)* warm water gushed out uncontrolled and under its own pressure into a concrete trough, overflowing to form the huge and wasteful pool at which hundreds of livestock drank and paddled. Stretching away from the edge of this pool the ground was dry and baked hard, compressed by all the hooves, and devoid of all vegetation except for a few desolate and damaged trees.

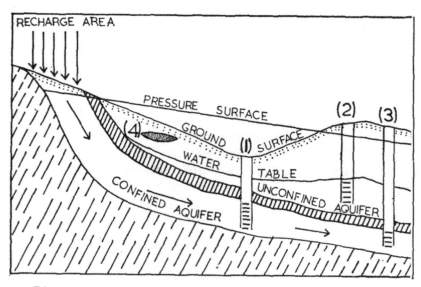

*Diagram 7. Types of groundwater aquifer: (1) artesian well. (2) water-table well. (3) sub-artesian well. (4) perched aquifer. Shaded areas indicate impermeable strata.*

The exploration of the groundwater system was begun in 1955 on the Nigerian side of the lake in order to demarcate economically useful sub-artesian and artesian water supplies for domestic, pastoral and agricultural[15] use.

**Water re-charge and losses.**

As explained earlier, the *Chari River is the recharge mainstay* of Lake Chad, pouring about 40.4 milliard m$^3$ of silt- and salt-laden water annually into the southern basin.

Some *rain* falls during the rainy season (April to July) on the lake surface but it is insignificant and unreliable compared with the recharge from the River Chari. Indeed we several times remarked the approach of heavy rain clouds, with visible heavy precipitation apparently 'hanging' below them, which never actually fell as far as the surface of the lake, but seemed to have dried up in mid air!

Moreover, a rise in water level at one point on the land surface by inundation waters may well result in a *sub-surface backflow* through permeable surface formations to landlocked pans and surface depressions producing a lagoonal inundation in otherwise isolated places.

**The lake has no surface outlet so where does all the water go?**

First, the greatest losses are by *evaporation*. Day temperatures are very high and the prevailing strong wind

## Chapter 4 - The puzzle of the puddle

from the north-east, the *Harmattan,* together cause very rapid surface evaporation. Water also *percolates into the sand* during the inundations and the heavy losses through the abundant floating vegetation by *evapo-transpiration* account for the remainder of natural losses. The significance of *additional losses* due to modern developments in irrigation for farming and the installation of those deep bore-holes will be discussed further in chapter 10 and 12..

While it is almost certain that water losses from the lake by inundation and infiltration of the Bahr el Ghazal and northern and eastern shores account for part of the 3 per cent of water lost by means other than evaporation, it is also probably true that a small proportion is lost to the western shores by inundation and by percolation following *local* rises in lake level occasioned during storms and gales. In the storm that we witnessed soon after the launching of the *Jolly Hippo* (See Chapter 2) the level of the lake rose about 30cm at the shore, and advanced about 100m in some places. This water was apparently all immediately absorbed into the dry sand, we saw less backflow towards the lake again than might be expected, for example, as after a tidal wave.

Even higher storm rises in level (others have reported advances of up to 500m) undoubtedly occur at intervals throughout the rainy season and must account for considerable water loss, probably not compensated for by

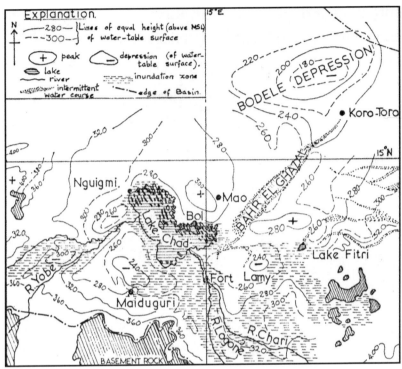

Map 10. Inundation zones and surface contours of the water table within the Chad Basin in the vicinity of Lake Chad.

the actual quantity of rain falling at the time. These storms do not usually last long and immediately the clouds have passed, the sun shines again with all its usual desiccating power, rapidly causing any surface moisture on the littoral to evaporate. Lunar tides occur, but their influence on the level of the lake surface is so small as to be negligible.

The morphology of the lake is not uniform and is constantly changing with the rise and fall in surface level. Indeed it is the complex cyclic oscillation of surface level of Lake Chad which determines the unique character not only of the lake itself as an environment but also of all the human activities associated with it.

While we may list the main facts as far as they are known today, and some of the biological and economic repercussions resulting from its queer hydrology, no one as yet knows for certain either what factors really control the hydrology of Lake Chad or what effects modern man's attempts at lake management, politely known as 'development programmes', may ultimately have upon it.

**3. The lake has no outlet, yet the open waters remains sweet.**

What is also so remarkable is that if Lake Chad suffers from prolonged drought, it should be expected to form a salt pan, like Lake Utah in the United States of America, or Lake Eyre in Australia. Lake Chad, however, contrary to normal lake behaviour, has not yet done this and the open water has a sufficiently low salinity at all times to be drinkable.

Extensive studies on lake salinity have been made both by scientists working at ORSTOM and at Malamfatori. The concentration of salts dissolved in the waters of the lake at any given time depends upon the balance on the one hand of water intake into the lake from rainfall and rivers, together with their content of dissolved salts, and from deposition from the winds, and on the other hand upon water losses by evaporation so concentrating the residue (evaporated water is always salt-free), by percolation (together with dissolved salts) into the lake shores and bed, and by salt intake into plants and animals.

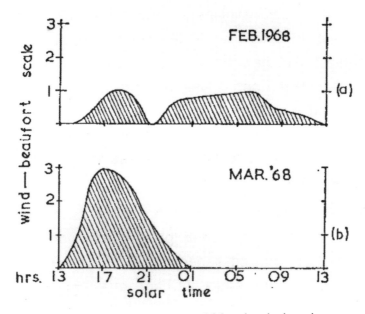

Diagram 8. February and March wind regimes.

An estimated 300kg/km$^2$ of dissolved salts (mainly chlorides and sulphates of marine origin) are precipitated in *rainwater,* on the lake surface. The mean annual tonnage of salts and silt brought to the lake in the waters of the *Chari* and *Logone rivers* is estimated at about 2 million metric tons. The influx of water from the Chari reaches its peak between October and December, so that a mass of fresh turbid water with its maximum load of silt and salts thrusts into the southern basin then.

## Chapter 4 - The puzzle of the puddle

During the period February to May, conditions become more stable, accompanied by a gradual overall rise in conductivity (salinity) in the vicinity of the region of the Great Barrier which now divides high salinities in the north basin from low salinities in the south basin. Actual deposition of salts then in the northern basin and the northern part of the southern basin at such times becomes very high indeed and is well exploited where it settles in shallow pans as 'natron' (an impure carbonate of sodium and magnesium locally and incorrectly called 'potash') and dries out as lumps and flat slabs. These are dug up by the local people. *(See Plate 22.4).*

This is not to be confused with a cooking salt extracted by the Shuwa and Kanuri people from various lake grasses locally called pagam (which is said to be salt-bearing only in the vicinity of Chad), *kalasilum* and *kanido* (which are said only to occur at Lake Chad), and from the saltbush or *siwak* tree *(Capparis aphilla)* called *kigu* in Kanuri, and *syak* in Shuwa-arabic. The branches of a Mimosa are also said to be used.

A well is dug, and a big pot put in the bottom. Then the grasses, or the leafy branches of the salt-bearing tree, are burnt and the ashes placed in a basket over the pot. Water is poured over them so that it filters into the pot, which is then boiled to dryness leaving a sediment of salt in the bottom. This contains potassium and sodium chlorides, although the major part consists of carbonate of lime.

The pot is then broken away from the residue of salt and used for animal husbandry and domestic purposes

Salts are also brought into the lake by the *winds* but only comprise about 1% of the total sediments deposited on the lake surface, the remainder consisting of quartz/diatomite grains. These salts mainly comprise calcium, magnesium, sodium, potassium, chloride, sulphate and carbonate radicals. The total deposition of wind-borne sediments on the lake surface is estimated at about 100 metric tons per square kilometre per annum. Most of this is brought by the *Harmattan* wind from the north-east and by the dust storms that precede tornadoes at certain seasons contributing not only to salinity but also to the decreasing lake depth.

### So how does the open water remain sweet?

In view of the existence of lake plants possessing such a high salt content, it seems very reasonable to suspect that the floating aquatic vegetation of Lake Chad plays a major role in salinity regulation in the lake. When the lake level is high and the volume of water greater, salt concentration in the open water is reduced and at the same time vegetational cover also becomes minimal. During extreme recession of the lake ('Little Chad') the aquatic vegetation has been seen to increase enormously in extent. It may be supposed that the increase in vegetation at this time is sufficient to account for the removal of a greatly increased amount of the salts dissolved in the lake waters. So salt concentrations are kept within the open lake waters at their normal mean. However, in 1905-8 the channels in the archipelagos became too brackish for man or beast to drink. Thus, at that time of excessive desiccation, apparently neither the salt losses by inundation and infiltration, nor by vegetational uptake, were adequate to balance the concentration increase resulting from evaporation from such a reduced volume of water.

The main seasonal changes affecting Lake Chad are typical of the northern part of Nigeria, with a cold, dry season lasting from November to early March, a long hot season from late March to June or July, followed by a brief rainy season in July and August, and a short hot season in September and October.

Shade temperatures on the lake tend to be lower than on the shores (due to the cooling effect of the wind) and humidity higher. Records indicate that, as far as rainfall is concerned, a narrow coastal belt near Malamfatori is the driest place in Nigeria with a mean rainfall of only 214mm, whereas at Yau, only 12.8km inland from this place, mean rainfall is about 304mm.

## PLATE 13
## YEDINA LIFE (A)

The *kadai*, or papyrus reed boat, is the traditional work boat of the Yedina people. It costs them nothing to build as papyrus and binding cord are made from plants native to the lake; it can be made in any size up to about about 6m long and 1.5m wide; it can carry quite heavy loads (such as several cattle together); it is unsinkable and serviceable (if kept on the water and never drawn up on shore) for at least two years; it is readily repaired if damaged, and can be safely paddled far out across the open waters even In a light breeze. It is not, however, readily manoeuvrable in high seas, but can keep afloat.

# PLATE 14
## THE YEDINA PEOPLE (B)
### *BUILDING A KADAI*

*(above left to right):* 1. Cutting papyrus for kadai building.
2. Mohammedu shows a complete papyrus plant.
3. Collecting and sorting fresh papyrus fronds for makina a kadai.

4 & 5. *(above).* Knotting the strong fibre ropes which are made from doum palm or grass fibres.
6. *(below left)* A small kadai is complete.

7. *(above)* Kadai after 2 year's use.

# PLATE 15
# YEDINA LIFE (C)

1. *(above).* A dry season fishing camp on a floating *Phragmites* island.

2. *(above).* The family head welcomed us one evening to his floating fishing camp: his name was Nuhu *(Noah).* 1969.

3. *(above).* The camp women and children posed for a photo, clasping their transistor radio as it played hymns from the 'Radio ELWA' Christian Gospel program. *Note* the mosquito net hanging from the grass supports.

4. *(above).* The boys ate their meal seated on slashed *Phragmites* grass stems. Even in 1969 an imported enamel bowl had replaced the traditional woven fibre dish.

# PLATE 16
# YEDINA LIFE (D)

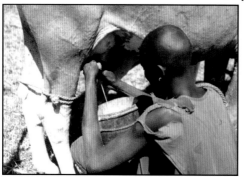

1. *(above).* Cattle are an essential part of Yedina life on the dune islands. The men usually do the milking, but the women rear the calves and make the woven fibre milk churns.

2. *(above)* Kourri calves are not left with the mothers all day but are given controlled suckling and then tethered at grass.

3. *(above).* Pure-bred *kourri* or Lake Chad bull with owner's brands on flanks.

4. *(above).* Pure *kourri* cow and calf. Both bulls and cows have the bulbous horns.

# PLATE 17
# YEDINA LIFE (E)

1. *(above).* Mixed breed cattle in the shallow muddy water in the drought of 1974 struggling, half-swimming, to another island in search of pasture.

2. *(above).* The scene as we approached a dune island: the Yedina with the their livestock. They always hurried over to welcome us.

3. *(above left).* Yedina hut on dune island: a thin-walled, light-weight papyrus/grass construction.
4. *(above right).* It is also portable, here seen being carried by hand in a circular movement towards the now distant shore, during the 1974 drought.

5. *(left).* In contrast, Kanuri huts are built only on the mainland, of a finer type of grass, and have much thicker walls and roof. They can be transported to and from the lake shore as necessitated by changing lake water levels, but are much heavier to lift.

# PLATE 18
# PEOPLE ON AND AROUND LAKE CHAD (A)

1. *(above).* Musician at Kukawa.  2. *(above).* Yedina boy musician on dune island.

3 to 5. *(above left to right).* Mounted Hausa piper, horn blower and drummer at the Muslim *Greater Beirom* festival at Kukawa.

6 and 7. *(right).* The District Head of Kukawa, Alhaji Abba Sadiq, with his family, and wearing his official ceremonial robes at the *Greater Beirom Festival, Id-el-Kabir,* 17th Feb.1970: he gave great support to our expedition to Lake Chad, helping and befriending us throughout our eight-month stay at the Lake.

# PLATE 19
# PEOPLE ON AND AROUND LAKE CHAD (B)
*PERSONALITIES WHOM WE MET DURING THE 1969/70 EXPEDITION.*

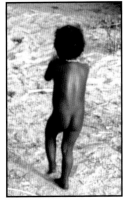  

1. *(above left).* Yedina: a fisherman of the future? 2. *(above centre).* No schooling yet for this island girl. 3. *(above right).* Schoolboy equipped with ink pot, Koranic text board, and school satchel.

4. *(above left).* Fulani lady.   5. *(above right).* Two Kanuri ladies

6. *(left).* A Hausa lady fills her clay pot with water drawn from a well.

7. *(right).* A Shuwa Arab lady rides her ox to market.

# PLATE 20
# PEOPLE ON AND AROUND LAKE CHAD (C)
*PERSONALITIES WE MET DURING THE 1969/70 EXPEDITION.*

1 and 2. *(above).* Yedina ladies on islands.   3. *(above).* Tuareg cameleer at Kukawa.

*(left to right).*
4. Fisherman.   5. General Trader.   6. Fulani camel trader.

*(left to right).*   7. Farmer.   8. Fisherman.   9. Young Fulani cameleer.

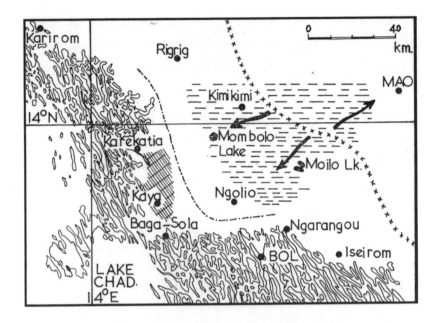

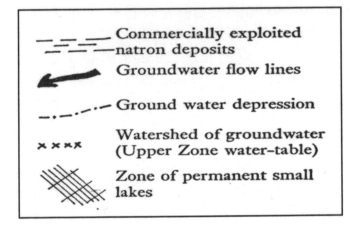

Map 11. Salt deposits on the north-eastern side of Lake Chad.

Apparently the presence of a wide expanse of water inhibits the formation of cumulo-nimbus cloud systems which tend to concentrate over land along the shore where thermal uplift is more pronounced. Storms developing from such cloud formations usually move in a westerly direction and precipitation takes place further inland. Trade winds blow strongly from the north-east on most mornings between mid- October and March, slackening by noon and backing north by late afternoon.

Winds are less predictable during the rest of the year, but in July, a strong west or south-westerly wind may blow from about 8.00 a.m. until noon, when it may change abruptly to an easterly wind. This morning wind is sometimes strong enough to result in a diurnal fall in lake level of as much as 25cm. This change in level caused by storm winds may also occur as a result of east and north-easterly winds and tornadoes. Our own experience of this at Portofino has already been described in Chapter 2.

## THE UNRESOLVED PUZZLE

Although so much methodical data collection and analysis has already been done, we are still no nearer to finding the solution to the real puzzle of the puddle; what causes and regulates the overall climatic cycle that determines the oscillatory regime within the whole Chad Basin and the lake itself.

- How can we predict the timing of its future major oscillations in terms of transgression or regression, inundation or desiccation?

- Dare we try to manage and manipulate the lake's water regime by building dykes or drainage installations?

- Would it be safe to clear papyrus fields, spray them with insecticides or use them for industrial purposes?

- What will happen if we exhaust, with our artesian wells, all the pressure water in the groundwater aquifers under the lake?

- And when the Chari River begins to add industrial effluents to its annual two million tons of dissolved natural salts and suspended silt, and when motor vessels and hovercraft crowd the shipping lanes across the lake, what will happen to the chemistry of the water, and the fishing industry that feeds a million protein-hungry people?     So......

*......the puzzle of the puddle remains as yet unresolved.*

*Chapter 4 - The puzzle of the puddle*

# PART II

# THE PLANTS AND ANIMALS OF LAKE CHAD

# Chapter 5

# The Flora

The vegetation of Lake Chad and its shores has been described both by early explorers and, more recently, by botanists working with the research organisations mentioned in Chapter 4. The type of vegetation characteristic of a given area depends primarily upon the climate, water supply and substrate and, in turn, sets the scene for the characteristic fauna.

**The vegetational zone.**

Lake Chad lies in the semi-arid zone of fixed dunes between the Sahara desert proper and the northern savannah belt of Africa. It thus falls mainly within the two savannah types known as 'Sahel' and 'Sudan', of which the former grades into semi-desert and the true Sahara desert to the north. Although the lake lies within this semi-arid zone, its own flora constitutes a distinct vegetational area with a characteristic aquatic and semi-aquatic range of plant species. It is not, however, particularly rich in its variety of phanerogamic (flowering) plants, although certain dominant species (e.g. papyrus) occur in great abundance.

Just prior to my 1969-70 expedition, a number of people who expressed interest in my plans were puzzled by the concept of a large lake which could not only exist in the southern part of the Sahara and contain hippopotami and floating islands, but even accommodate my yacht. The popular image of desert and semi-desert tends to be limited to Christmas-card style views of 'three wise men' mounted on three camels, silhouetted against a cloudless sky as they plod in single file along the summit of a sand dune devoid of rocks or vegetation. Alternatively, 'The Three' sit chatting beside an oasis consisting of a smallish pool, with a few date palms, kneeling camels and a navy blue, star-pierced sky.

Around the southern end of Lake Chad, Sudan savannah merges into Sahel. The Sudan savannah tends to consist of a fairly open type of woodland, with larger, broad-leafed trees such as *Piliostigma reticulata, Bauhinia rufescens, Ziziphus mauritiana* and *Adansonia digitata* (the baobab), but dominated by thorny acacias such as *Acacia seyal, A. tertilis, A. arabica* and *A. albida* or, in certain areas, by the doum palm, *Hyphaenae thebaica*.

The baobab is the 'elephant' of the plant world, its bole frequently exceeding 15m in circumference, but rarely attaining 15m in height. The branches are slender in proportion, with large white flowers which persist for only a brief period, to be superseded by large, pendulous, oval fruits about 35cm long by 15cm diameter. The leaves are

## Chapter 5 - The Flora

dark green in colour and digitate, about 8cm long. The bark is grey with a purple sheen, its shade varying with the season. It has strong callus-growing properties, and repairs injuries to its surface with great vigour. The outer bark is soft, spongy and full of sap, while the inner layers are fibrous and have an important use in rope-making *(Diagram 10, page 141).*

Almost every part of the baobab has one or more uses, and the bole frequently contains several gallons of potable water. In some areas near the lake, each baobab tree is traditionally owned by someone, and cannot be utilised or cut down without his permission. The leaves and fruit are harvested by slashing off the extremities of the branches leaving the tree looking somewhat grotesque and ill-proportioned, with a pollarded and shorn appearance. When the bark is harvested, strips and patches are cut from around the base of the trunk, leaving a slash which is at first mottled red and white and then becomes overgrown, nodular and grey in colour.

Baobabs are favoured as food by African elephants, but are rarely used by them excessively unless the area becomes overpopulated with elephants. They then tend to rip, prod and destroy baobab trees, sometimes pushing them right over, a feat rarely performed even by bulldozers. The scars made by elephants also acquire a nodular appearance generally distinguishable from man-made scars. The presence of baobabs scarred and damaged in this way by elephants thus serves as an indicator of elephant/habitat imbalance, and the degree to which it has progressed. Around Lake Chad, however, elephants have become so scarce that one rarely encounters evidence of damage done by them in recent years to the baobabs. It is commonly believed that baobabs and doum palms are distributed primarily by man or by elephants. They are thus indicative either of present or former human habitation, or of present or former elephant habitat and migration routes. I have failed to locate scientific data about this: it would be interesting to know how much truth there is in this theory.

The Kanuri and Hausa name for the baobab tree is *kuka* (pronounced 'koo-kah'). When the great Shehu Laminu founded the city that was to become the seat of the second Sefuwa Dynasty in 1814 he called it 'Kuka' after the baobab tree under the shade of which he sat as he read the final words of the Koran.* Today the town is generally known as 'Kukawa'. The original baobab still stands today to the west of the town of Kukawa.

In the rainy season, Sudan savannah becomes a rich green in colour, with a thick ground cover of grasses and other herbaceous plants; whereas in the dry season most of the trees lose their leaves for several weeks and the ground cover dries out leaving only a sparse, parched, khaki or straw-coloured remnant over cracked clay.

---

* See Chapter 7

Travelling northwards, a transitional zone is passed consisting of low scrub grading to the true Sahel, characterised by its flat topped *Acacia raddiana* trees, some gnarled, low *Commiphora africana* trees, *Hyphaene thebaica* (doum palms) and low bushes called *Leptadenia pyrotechnica*. A bluish-green, round-leaved shrub is very common in this zone. It bears large green fruits the size of a grapefruit which are a favourite food of the red patas monkeys that frequent the savannah around the lake: this plant is *Calotropis procera* (Sodom apple). *(See Plate 21.6).*

When I first arrived at the lake in mid-August 1969, the vegetation between Maiduguri and Baga was already green but the grass not yet tall. By September, however, it had become so dense and tall that it was in many places almost impenetrable – guinea-fowl and greater and lesser bustards simply vanishing into it. The trees at this time provided wide pools of shade, flies were abundant, and the livestock were becoming fat and contented-looking. One feature that could not be overlooked during September and October was the incredible abundance of dung beetles, many of which were relatively enormous beasts of up to four or five centimetres in length, which flew heavily and laboriously like miniature transport 'planes

*Diagram 9: Baobab trees in the dry season, and detail of the flower, leaves and fruit.*

They were to be seen wherever the livestock had passed by, busily rolling dung into bigger and bigger balls, some of these attaining a size comparable to that of a tennis ball. They pushed these rapidly backwards with their third pair of legs towards suitably soft burial sites. Of course the number of these beetles was particularly high along the main road and its verges where the concentration of dung, left by the hundreds of cattle, sheep, goats, donkeys, horses and camels as they were taken to and from the artesian overspill pools, was also highest.

As December moved on into January, the vegetation turned first khaki, as the grass seeds ripened and fell, and then straw-coloured as the trees lost their leaves and the grass stems dried out and broke up. On the flat inundation plains along the south-western shores, the perennial grasses turned almost white or whitish-grey in colour. These are mainly *Pennisetum pedicellatum,* and some *Sporobolus spicatus*; but very large areas of the vicious hooked grass *Cenchrus biflorus* (cramcram) also occur here, and in late November and December were at their most vicious and impenetrable stage.

By the end of February, the livestock were invading every corner of the savannah in their desperate search for a few straws to munch and thin tokens of shade in which to pretend to shelter from the sun. They would crowd in groups of any number from three to thirty animals into the pathetic, trellis-like patches of shade cast by the twigs and branches of leafless trees.

At this time too, the herdsmen and shepherds slashed the upper branches of the acacias, of which the leaf buds were now beginning to sprout in anticipation of the approaching rains, to bring them within reach of their stock - a practice most detrimental to these trees. Even by the end of February much of the exposed clayey soil had been covered by layers of fine, white sand, blown into ripples by the *Harmattan* wind.

The *Harmattan* now arrived from the north-east laden with its heaviest burden of fine diatomaceous sand-dust, which it left hanging in the atmosphere like a thick dry mist over the land, to settle gradually as the wind subsided. The countryside seemed desiccated and thirsty, and life for man and beast centred constantly from dawn to dusk on the business of drawing water, scratching for the few remaining grass straws and edible roots, and searching to find even an 'apology' for shade. The sand-dust filled eyes, nose, ears and throat, and the cattle, sheep and horses coughed and wheezed, while an epidemic (an annual event at this time of year) of cerebro-spinal meningitis swept through the populace, taking its toll of health and life. Only the goats and camels continued to look fat and fit.

Further north, on the dunes of the Sahel around Malamfatori, Nguigmi and the north-eastern shores, the green colour of the grass was much more fleeting, quickly replaced by the glaring bleached appearance of the white sand among the parched tufts of dune brush and grasses.

As we drove the Landrover during the expedition along the north-western shores, August's battles with alternating sticky mud of the clay pans and insecure sandy dunes changed in February to continuous struggles with deep, soft blown sand; while on the south-eastern side, the soft, saturated dark mud of the *firki* plains, first flooded by rivers and then by lake inundation, gave way by February to a dry, cracked, rough-cast surface that so shakes the vehicle that some people say it makes even their artificial dentures ache when crossing it.

Even during the driest part of the year, certain species of trees bear leaves - for example, doum palms - and these stand out in refreshing contrast to the otherwise barren appearance of the land. Near the lake shore, the papyrus forms a green line, but even this acquires a brownish look as the flowers mature, and tends to lose its clarity as the midday mirages lift the line somewhere up into the sky leaving an indeterminate scintillating gap between it and where the land ends.

**The lake flora.**

The vegetation of the lake itself falls naturally into six ecological categories or associations:[16,17,18,19],

*(1) marginal vegetation* - plants growing at the edge of the lake in areas subject to flooding.

*(2) swamp vegetation* - plants standing in water but with their upper parts frequently floating clear of the substrate, and occasionally temporarily free-floating.

*(3) free-floating vegetation* - plants that grow on the water surface but are never rooted.

*(4) epiphytic vegetation* - plants growing on rooted or free-floating rafts of swamp vegetation.

*(5) submerged rooted vegetation* - waterweeds.

*(6) phytoplankton* -microvegetation freely floating in the water.

These plant associations are influenced by several environmental factors of which the most important are:

*(1) Exposure.* Generally the flora is richer the more sheltered is the locality. Thus on a transect running due west from the eastern shore of the lake to the Yobe mouth, a progressive decrease in species is noticeable as the islands become more and more exposed to the winds and wave action of the open waters. On the most exposed, most westerly, islands to be found along this transect line, only one persistent species of the higher plants occurs, namely the reed-grass *Phragmites australis.*

*(2) Changes in lake level.* Both seasonal and long-term changes in surface level of the water are very significant indeed in determining both the quantity and distribution of the aquatic flora. As mentioned earlier, a rise in lake level results in an increase in detached, vagrant islands of floating vegetation, or kirtas, and, in the long term, a decrease in the overall quantity of aquatic vegetation. In contrast, a fall in lake level increases both the proportion and the actual quantity (measurable as biomass, or weight per land or water unit area) of rooted vegetation, as well as resulting in colonisation of freshly exposed shores by terrestrial plants.

*(3) The nature of the sediments.* The character of the bottom deposits, whether sand or clay, influences plant

distribution in the lake, Nile water lettuce *(Pistia stratiotes)* for example, developing more abundantly over the clay sediments of the El Beïd and Yobe deltas, than over sand.

*(4) The physico-chemical composition of the water.* The seasonal salinity gradients and swirls of the north and south pools, as well as the overall increase in actual salinity from the Chari delta to the northern shores are described in Chapter 4. Analyses of the distribution of aquatic plant species in the lake are found to conform closely to the salinity patterns. This may be illustrated by the progressive disappearance from south to north of the aquatic grass *Vossia,* and of papyrus, while reedmace becomes completely replaced by the reed-grass, *Phragmites.* Finally, as the northern shores and natron pans, swamps and pools of Kanem are reached, *Cyperus laevigatus,* a rush confined to swamps characterised by heavy natron deposits, appears and extends in comparable situations right across the Sahara to just south of Libya.

**The micro-fauna or phytoplankton**

The micro-flora of the water, also called phytoplankton, is of vital importance to the life of the micro-fauna or zooplankton, other invertebrates and some kinds of vertebrate aquatic fauna.

Plankton tend to make diurnal movements up and down in the water following gradients of light intensity, concentrations of dissolved gases, salinity and temperature.

The phytoplankton of Lake Chad[20], tends to be dominated by comparatively large, colonial blue-green algae of the genus *Microcystis,* which in some areas comprise 9 per cent of all the open water phytoplankton. Diatoms exhibit great species diversity among the dune islands of the archipelagos, the commonest species being one called *Melosira granulata.* Among the desmids, *Closterium* is the commonest genus, while several species of *euglenoids* and *chrysophysids* occur in the sheltered areas such as the mouth of the Yobe. So far about eighty-five species, representing sixty-eight genera of phytoplankton, have been distinguished in Lake Chad, but many more remain as yet unidentified.

**The macro-flora of the lake.**

The macro-flora[21] (that is all plants not classed as phytoplankton) of the lake falls into eight categories according to location. These are:

*(i) Temporary swamps in the Chari and El Beïd basins* and the temporary courses of their distributaries. These have a herbaceous, ephemeral type of vegetation, which contrasts with that of the banks of the Chari River which

are lined with low stands of the willow, *Salix ledermannii*, and a mimosa shrub, *Mimosa pigra*.

(ii) *Sands of the Chari delta:* here the vegetation consists primarily of sand-loving herbs and various shrubs.

(iii) *Silty areas at the mouths of the El Beïd and Yobe rivers:* a characteristic association of plants such as water lettuces *(Pistia),* and white and blue water lilies *(Nymphaea lotus* and *N. micrantha)*, waterweeds *(Potamogeton schweinfurthii, Vallisneria aethiopica)* and semi-aquatic plants typified by genera such as *Ludwigia*.

(iv) *Deltas of the Chari and El Beïd rivers:* aquatic meadows and floating rafts of the grass, *Vossia cuspidata*, with sporadic occurrences of papyrus *(Cyperus papyrus)* and the reed-grass, *Phragmites australis*.

(v) *The open waters:* poor in vegetation except in the clearer waters of the north basin where submerged vegetation may overlie bench islands. The waters of the north basin contain much less suspended silt than those of the south basin, and in some places are very clear indeed.

This difference between the north and south basins is, of course, directly related to the influence of the Chari River, discharging its tremendous burden of suspended silt into the lake within the south basin. This is circulated by wind action and swirl throughout the waters of the south basin so that they always look silty, and, on standing, deposit a considerable quantity of sediment. This suspended silt, however, does not make the water unpleasant to drink and, in fact, the waters of the south basin, untreated, taste very pleasant and refreshing. It is due to this heavy burden of silt that the lake is apparently getting progressively shallower - i.e. by a gradual raising of the level of the bottom. The silt discharged into the lake is greatly increased also by the deposition on the surface of wind-borne sediments amounting to some 110 metric tons per square kilometre per annum. Strangely enough, no fixed scales for recording changes in bottom level at divers points in the lake have yet been established, although it would seem to be an obviously interesting thing to do.*

*(vi) Bench islands:* typically covered by meadows and floating rafts of the grass *Vossia cuspidata,* papyrus *(Cyperus papyrus),* the reed grass *Phragmites* and the reedmace, *Typha domingensis.* The floating vegetational cover of some bench islands frequently contains a clearer area of water towards the centre with scattered bulrushes and water lilies. These clearer areas are often the favoured residences of hippopotami, and their regular paths in and out from the open waters outside are generally well-defined as they pass through the *Phragmites,* leaving the floating stems thrust aside in rather muddy disarray. In the case of much-used hippo

---

* Now (2002) in operation in the southern basin.

trails, the edges are tidier in appearance, sometimes bordered by margins of creeping ferns and other low epiphytes.

Moving from south to north, the arrangement (generally centripetal) of plants on each bench island changes : in the south the order is *Vossia* - papyrus - *Phragmites;* further north *C. papyrus - Phragmites - Typha;* and in the extreme north *Phragmites - Typha* (with no papyrus).

It is on these rafts of floating *Vossia,* papyrus, reed-grasses and reedmace, that a complex association of epiphytic plants may occur, the floating rhizomes, stolons (horizontal stems) and broken stems forming a mat on which sand and plant debris accumulate as a substrate in which the epiphytes take root. These include a variety of forms, mainly climbers, creepers and ferns, such as the flowering *Commelina diffusa, Ipomoea rubens* (a convolvulus), *Luffa cylindricus, Murkia maderaspatana, Oxystelma bornuense, Vigna spp,* and the creeping ferns *Cyclosorus gongylodes* and *Thelypteris totta.* Unfortunately few of these plant species have english names.

(vii) *The islands of the archipelagos:* characteristically surrounded by a floating fringe of papyrus and reeds; then a zone of inundation, often rich in such plants as *Nymphaea* (water lilies), *Potamageton schweinfurthii* (pondweed), *Utricularia sp* (bladderwort), clumps of *Cyperus mundii,* and meadows of *Leersia hexandra* (an excellent cattle pasture grass); and finally a central sandy zone, generally dry with the normal flood limit marked by a belt of doum palms *(Hyphaene thebaica)* and various other shrubs and low trees such as herbaceous cassias, the convolvulus vine, *Ipomoea rubens* with its distinctive mauve flowers, and *Sporobolus* grass.

The south-easterly archipelago characteristically has flat low dune islands with the usual fringe of floating vegetation and the occasional relict stand of ambach *(Herminiera elaphroxylon\*)*. The ambach is a leguminous tree of which the density is half that of cork, with the trunk occasionally attaining a diameter of 40cm and a height of 8 metres. More commonly, however, it is about 15cm diameter and 5m high. Ambach occurs only in a very humid environment, among reeds, and rooted in not more than 2m depth of water. These trees were very numerous in 1908 throughout the course followed by General Tilho through the archipelagos, and were described on the banks of the islands in the vicinity of Bol in 1949, with a veritable forest of them between Kindin and Boesmerom. Almost all of these disappeared during the exceptional rise in lake level of 1955-6, and today (1969/70) ambach trees on Lake Chad are extremely rare. *(See Plate 25.2)*.

The swamp fringe of these dune islands is followed by a very extensive inundation zone with meadows of the

---

\* Also named *Aeschynomene elaphroxylon*.

pasture grasses *Leersia hexandra* and *Echinochloa pyramidalis*. Further north, the shoreline fringes consist of extensive, usually dry, meadows of the grasses *Panicum repens* and *Sporobolus spicatus,* and are frequently the sites of semi-permanent Yedina villages. These islands are generally devoid of trees, but towards the north-east, near Baga Sola, stands of doum palms crown these islands above the normal flood limit

Further north again, higher dune islands are found. These are never nowadays inundated, and carry savannah woodland with the trees and shrubs *Acacia seyal, Balanites aegyptiaca* (desert date), *Calotropis procera* (Sodom apple), the broom-like *Leptadenia,* and a persistent green herbaceous ground cover dependent at all seasons on the groundwater table. On other islands with a similar relief but a deeper water table, typical sahel savannah occurs, green in the brief rainy season, and otherwise a parched, sandy dry colour.

*(viii) To the north and east of Lake Chad lie the innumerable inter-dunary pools, pans and marshes of the dune field of Kanem.* These are mainly maintained by meteoric water (rain) and ground water, but their water level frequently shows a relationship to that of Lake Chad. The Bahr el Ghazal *(See pages 92-93)* is part of this system, but it appears to have an even more direct relationship to Lake Chad than the smaller pools, pans and marshes.

The latter are characterised by their high salinity due to the presence of heavy deposits of natron. As may be expected, the flora of this area is much poorer than that of the lake, and is restricted to salt-tolerant and salt-loving plants. Three main vegetational types are found here.

These are: (a) marshes with islands of *Phragmites* reed-grass intermingled with vines, a shore belt of the grass *Panicum repens,* and then clumps of *Sporobolus spicatus* grass outside this on the sand dunes; (b) pools with sickly looking clumps of *Phragmites* reed-grass, the rush *Cyperus laevigatus* with *Panicum repens* grass around the shore, and then *Sporobolus spicatus* clumps outside on the dunes; (c) marshes devoid of aquatic vegetation but with a deep green marginal fringe of the rush *Cyperus laevigacus* and then the usual clumps of *Sporobolus spicatus* grass on the sand beyond.

**The flora and salinity.**

Of all the factors influencing this south-north gradient of vegetation, salinity appears to be the primary determinant, with its minimum near the Chari River mouth and its maximum along the north-eastern lake shores and the natron pans of Kanem. Localised variations within the overall pattern are influenced by seasonal salinity changes and swirl, exposure to wind and waves, the nature of the substrate and changes in lake level.

The more one ponders the factors regulating vegetational balance and considers, in particular, those controlling salinity levels, vegetational distribution and biomass, one realises with awe - and anxiety - how delicately poised is this fascinating and fragile ecosystem.

*This then is the environment not only of the natural fauna, including the micro-fauna (zooplankton) of the water, the amphibious macro-fauna, the fish, birds, reptiles and mammals of the shores, but also of the people and their livestock, whose whole history and modern way of life centres in and around Lake Chad.*

# Chapter 6

# The Fauna

The fauna of Lake Chad spans the whole size range of living animals on the world's continents today - from the minute aquatic micro-fauna (zooplankton) to the largest of African elephants.

**The micro-fauna or plankton.**

Zooplankton[22] comprises those very minute creatures that live with, and many of them on, the phytoplankton of any suitable aquatic environment, sharing the diurnal up and down movements of the plants within the water.

In Lake Chad, the zooplankton includes members of the *HYDROZOA* (better known as hydras and anemones) of which a freshwater medusa of the genus Limnocnida has been found; the *ROTIFERA* (or rotifers), of which thirty-three species have been distinguished up to date, *CRUSTACEA* (of which the larger members are generally more familiar as crabs, prawns and suchlike), of which fourteen kinds of Cladocera, two kinds of Copepods, and at least four species of Cyclopoida have been recognised. Many of these are also known to belong to the micro-faunas of Egyptian, Nigerian, Tchadian and East African rivers and lakes. Actual plankton counts using a small type of plankton net (mouth about 30cm diameter) were in the range 30,000-137,000 individuals per cubic metre of water, while densities estimated from catches using a bigger and more sophisticated net of comparable mesh suggested densities of between 90,000 and 400,000 individuals per cubic metre of water. If the latter is the figure nearer to the true plankton density of Lake Chad, then it is the richest of all African lakes in terms of its standing crop of plankton. The significance of this lies in the fact that plankton is of vital importance in determining the size and quality of the fish population, in turn a factor of vital importance to the welfare of the lake people.

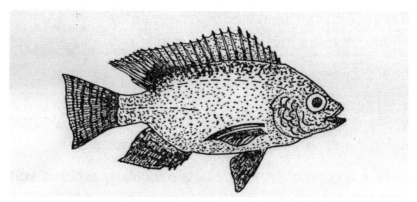

Diagram 10. Tilapia nilotica. In Lake Chad, the tilapias occur most abundantly in the more heavily vegetated zones.

**Benthos**

The fauna of the lake bottom is called benthos, and in Lake Chad its distribution is directly related to that of the various types of bottom sediments described in Chapter 4.[23]

For example, soft mud which is free of detritus, or plant debris, always contains an appreciable quantity of dead shells of the mollusc *Melania tuberculata* in densities of up to 6,000 individuals per square metre of lake floor. Fine sand deposits invariably contain the dead shells of the mollusc *Corbicula africana.*

Much of the mud, however, contains a proportion of plant detritus. Towards the western shores of the north pool, the proportion of this vegetable matter is low, but rises towards the north and east as the outer fringes of the reed islands are approached, and is very high within the northern and eastern channels of the archipelago. Much of it is derived from the reed *Phragmites australis,* and further south from fragments of the roots and rhizomes of papyrus.

The two most important factors correlating with benthos distribution on Lake Chad are the proportion of plant detritus and particle size in the bottom deposit.

Five groups of benthic animals regularly occur in soft mud: oligochaetes (worms), ostracods (small crustaceans), chironomids (insects), and two kinds of gastropods (molluscs), namely *Bellamya unicolor* and *Melania tuberculata.* All of these, except ostracods, are substrate feeders and prefer a fine homogeneous bottom with a low detritus content. Ostracods, however, are common in mud rich in detritus and rarely occur in substrates poor in plant matter. Relatively few organisms occur on deposits of hard clay. In a survey carried out in the south basin, *Cleopatra cyclosomoides* turned out to be the commonest mollusc on all types of substrate including clays, mud, peat and sand near Bol, where all these contain a high proportion of plant detritus. In contrast, the mollusc *Mellania tuberculata* is rare around Bol, but is common in the open waters of the south pool.

It is impossible, without becoming too technical, to give any real idea of the full range of either zooplankton or the benthic animals living in Lake Chad. A single example must serve to illustrate the range: a French scientist[24] has already recorded from Lake Chad no less than seventy species, all belonging to the single family of flies known as chironomids. There are also, of course, innumerable other flies in addition belonging to other families.

Biomass is a useful measure of living matter (plant and/or animal) in a given habitat.[25] It is the weight of living material per unit area of the substrate. In Lake Chad, total mollusc biomass has been calculated at sample stations with the following results:

(i)   On mud, near Malamfatori     1,038kg/hectare
(ii)   On mud, near Bol     18kg/hectare
(iii)   On sand, near Malamfatori     168kg/hectare
(iv)   On sand, near Bol     25kg/hectare
(v)   On soft clay, near Bol     540kg/hectare

## The surface fauna

On the 1969-70 expedition, no attempt was made to study either the invertebrate fauna or the amphibia and reptiles. These fields were too vast, and would call for prolonged and concentrated study. We noticed innumerable dragonflies *(ODONATA),* two-winged flies including gnats and mosquitoes *(DIPTERA)* and bees, wasps and ants *(HYMENOPTERA).* There were numerous water beetles, including the ever-amusing whirligig beetles *(Gyrinidae)* swimming their continuous circles on the water surface. Fireflies and glow-worms illuminated the darkness of the night: both are also kinds of beetle *(COLEOPTERA).* Of the *LEPIDOPTERA,* small moths used to come to the navigation and cabin lights at night, and an occasional butterfly sometimes flapped across the sun drenched lagoons, sometimes honouring us by taking a brief rest on the yacht. On the reeds and other vegetation of the floating islands, we saw a few plant bugs *(HEMIPTERA),* while crickets and their allies *(ORTHOPTERA)* kept up a vigorous trilling chorus, day and night and earwigs invaded the yacht when it lay at anchor.

The desert locust, so well known for its swarming and migratory propensities, sometimes visits the Chad basin but, due to international controls, none of the overwhelming invasions so common in the past has occurred for many years now.*

---

* Swarms, during the period 1972-2002, have been recorded in increasing numbers

## Chapter 6 - The Fauna

We made no attempt to study spiders, although gossamer strands sometimes wafted into the rigging as we passed close by floating islands, or hung, dew-spangled between the papyrus and reed fronds in the early morning.

Every night the tree frogs held their chorus of twitters and chirrups, and it was tempting to try to collect them and identify them, but our facilities and available space aboard for making large collections of preserved specimens of animals were limited, and I resisted all my natural collector's inclinations in this respect.

### Fish

The major divisions of the lake (archipelagos, bench island zones, open waters and swamps) represent well-defined habitat types in relation to fish distribution and populations.

The fish of Lake Chad show affinities with those both of the Nile and the Benue-Niger River systems[26]. Evidence of a past connection between the Chad Basin and these systems provides an explanation for these affinities, but, whereas there seems to have been a two-way connection between the Nile and Chad Basin as regards the aquatic fauna, there appears to have been only a one-way connection between the Chad and Benue-Niger Basins, namely from the Chad Basin into the Benue-Niger Basin.

This view is supported by the fact that the Citharinid fish, *Citharidium ansorgei* (a striped moonfish), is found *only below* the Gauthiot Falls (on the upper Benue) and does not occur at all above them. Moreover, the view that this connection is unilateral is strengthened by the fact that the West African manatee, *Trichechus senegalensis* (a large, herbivorous, entirely aquatic mammal) occurs in the Benue-Niger system, but has never been recorded at any time within the Chari-Logone-Chad system, nor in any other rivers or pools within either the Chad or Nile Basins. It is relevant in this connection that the manatee, unlike the hippopotamus, is incapable of movement, or even of temporary survival, on land, and is therefore incapable of any kind of terrestrial or amphibious migration between separate bodies of water

As regards fish distribution and population abundance within Lake Chad, the archipelagos far outstrip all other parts of the lake. In these zones, however, particular areas were, at certain periods of reduced lake level in the past, so heavily over-populated with fish that clear signs of degeneration were manifest, shown in such features as malformed scales, skeletal abnormalities and unusually heavy parasitic infestations. In these areas also, where water weed is abundant, large numbers of the smaller species of fish occur, such as barbels *(Barbel leonensis, B. pleurophlyxis, B callipterus), silversides (Alestes dageti, Micralestes sp, Haplochromis and Paradistichodus sp)*, together with juveniles of the larger species. *Hydrocyon forskali* is a fish particularly characteristic of the southern basin of the lake, while tilapias *(Diag. 10)* belong essentially to the heavily vegetated parts. The deltas of the Chari and Yobe rivers also yield a good fish harvest at all times; and many juveniles of the fish which breed in the Yobe are trapped as the river waters decrease in February.

Quantitatively, the best represented families occur in Lake Chad in the following order:

| | |
|---|---|
| *CHARACIDAE* | tiger fishes, micralestes, silversides, African pike, grass-eaters, striped moonfishes, etc. |
| *SILURIDAE* | mudfishes, butterfishes, flagfins, and catfishes. |
| *MORMYRIDA* | trunkfishes *(Diag.12, page 165)*, stoneheads, mormyrids. |
| *CICHLIDAE* | jewel fish, tilapias. |
| *CYPRINIDAE* | African carp, barbels. |
| *CYPRINODONTIDAE* | toothed carp, killifishes. |

**Reptiles**

In the lagoons water tortoises or terrapins are numerous, but I never went out of my way to try to capture any. As the boats nosed through the still waters, a terrapin's head might be hastily withdrawn with a slight swirl, defying all attempts to make a successful grab by hand. Snakes and monitor lizards are also common in the lagoons, and we would sometimes see a snake cleaving a V-shaped trail along the surface. Fishermen not infrequently catch snakes in their fishing nets, the skins of pythons (which are numerous in the lake and its fringe swamps) being traded profitably for use in making travel and hand bags, wallets and purses for the tourist trade. Swimming snakes usually attempt to climb aboard any substantial floating object like a vagrant island or a boat.

On one occasion, about 1.30p.m., I was at the helm of the *Jolly Hippo* as we were returning to Baga from a trip to a point offshore near Ngurno, when I noticed something resembling my idea of a 'Loch Ness monster'. It seemed to be very long and dark, and undulated along the surface of the water.

The sun was very hot and the heat haze over the surface of the lake vibrant and scintillating so that my monster really did look like some creature from the dark ages of myth and legend. I called out to my fellow crewmen, both of whom were sleepily reclining on the upper deck. Each in turn opened one eye, pronounced it a piece of floating papyrus, and snoozed off again. I was far too intrigued to accept this explanation and steered over towards it. Soon I saw it was an African rock python, not in fact a very large one, swimming deliberately on a westerly course straight across the 1.5km of choppy open water between two anchored floating islands.

Chapter 6 - The Fauna

Immediately the *Jolly Hippo* approached, it turned and tried to come aboard but, fearing that it might choose the exhaust line of the yacht as its entry point, I steered away from it.  We circled it for some time until it got somewhat disturbed and dived, remaining under for longish periods, on one occasion for nearly twenty minutes.  Finally, we decided to capture it into the *Sportyak* which had no awkward places where it could get itself stuck or entangled; and this was effected by one member of the crew diving into the water and grabbing it - not really a difficult feat as it was only, as we found on measuring it, 2.30 metres long.

*Diagram 11.  It was an african rock python.*

It was also very tired and did not struggle much as it was dropped into a nice damp sack. We called this snake 'Pythagoras' and it lived in captivity for sixteen months, when it died of ulcerative stomatitis, apparently contracted from another captive snake. In that period it grew about 30cms in length, and was known on more than one occasion to swallow no less than three adult guinea pigs in quick succession. 'Pythagoras', however, never became really tame as do some African rock pythons. Perhaps it resented confinement after its wild free life in a lake approximately 22,000 sq. km in area.

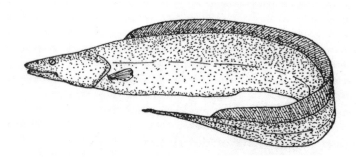

*Diagram 12. Trunkfish, Gymnarchus niloticus. In the Hausa language this fish is called Sarki (Chief) and is considered to be the most noble and most palatable of all the Chadian fish, often being used as a gift or tribute to a chief. We found it too strongly flavoured, however, for our taste.*

There are said to be three species of crocodile *(Crocodylus niloticus, C.calaphactus* and *Osteoloemus tetraspis)* in Lake Chad and, of these, the Nile crocodile apparently grew to a very large size in the past. The ancient skull of one was collected on the strandline south of Malamfatori recently. It was 60cm in length - i.e. belonged to a crocodile of total length approximately 4.2m. While we were at the lake, a number of juvenile crocodiles were sold to us, but unfortunately all these belonged to the species *Crocodylus niloticus* (the Nile crocodile), and we never saw either of the other two kinds. So few crocodiles remain in the lake today that we did not ever see a single wild one, although we were told that they are still comparatively common in certain parts of the archipelagos and the river delta areas. The fishermen catch small ones from time to time in their fish traps and nets.

Chameleons and geckos also occur, but we only encountered these on shore. Minetti housed a good population of geckos which lived on the walls and ceilings and helped to control the unbelievably abundant mosquitoes. Geckos may also occur on the floating vegetation of the lake, but we did not see any and no reference to any research findings about them could be located.

Monitor lizards usually only made their presence known to us when they launched themselves into the water from a bank of vegetation with a swirl or a 'plop'. It is the handsome black and yellow Nile monitor which occurs in the waters of Lake Chad. This species *(Varanus niloticus)* grows very large in some parts of Africa, but really large ones are now rare in Lake Chad due to their systematic destruction for the sake of their commercially valuable skins.

However, one day I encountered a large, very fat one, well over 2m in length. I was in the process of stalking a herd of elephants in a lagoon and had waded ashore through thick mud and into the damp, peaty, trembling material bearing the papyrus fringing the lagoon.

*Diagram 13. The herd of eight bull elephants that lived in the lagoons near Portofino and Baga during the 1969-70 expedition.*

The elephants were less than 20m from us, but we were hidden from them by the papyrus screen and the breeze was in our favour. Step by step we eased ourselves along the muddy trail they had just made up the bank.

## Chapter 6 - The Fauna

I was leading, followed by Mohammedu and a fisherman called Manu, when suddenly I realised that my right foot was about to tread straight on the back of this large, fat, sleeping monitor. My foot froze, rigidly poised in mid-air. Simultaneously, the monitor awoke, raised his head, took a prolonged look at my foot, and then waddled away past my companions down the trail into the water. In any other circumstances we should have attempted to photograph or capture it.

As it was, I was glad to have seen it in time to avoid treading on it as they are able both to bite viciously and to whip one brutally with the hard, horny tail. Like snake skins, monitor lizard skins are very popular in the countries around the lake for making articles for the tourist trade, and on one occasion, in Maiduguri, I was shown a warehouse inside the old city of Yerwa stacked from floor to ceiling with the crudely tanned skins of pythons, monitors and crocodiles from Lake Chad and its feeder rivers, as well as sitatunga (marsh antelope) hides and elephant tusks. There must have been thousands of skins in that room, and this was but one of many similar warehouses in Maiduguri.

On shore we saw few snakes, but one day, at Minetti, a funnel-type rodent trap was inadvertently left in the so-called garden near the house. Next morning we found that a small toad had hopped into it but, not having time to release it, we just pushed the trap into a shady place until our return later in the day. On our return, we found that a carpet viper *(Echis carinatus),* a small, very deadly sand viper, had entered the trap after the toad, which it had killed and eaten. Then it had attempted to leave through the wire mesh, only to find that its now distended stomach was too fat to pass through. It could not reverse either, due to the fact that this species has 'keeled' scales, and these acted as barbs holding it firmly fixed against the wire.

The following day, the road camp guard led us quietly to a little sandy hummock near the house and showed us another carpet viper sunning itself on the summit. Each of these specimens (which we duly collected as we did not care to have these very dangerous snakes so near the house) was just about 60cms in length.

### Birds

There is a very numerous avian fauna of Lake Chad, comprising not only a wide species range, but also residents as well as migrants.[27]

Approaching Lake Chad from the south-west, as we did in 1969, the traveller cannot help being astonished and thrilled by the incredible abundance of the exquisite carmine bee-eater *(Merops nubicus)* along the road. These birds are insect eaters and, since the installation of telegraph poles and wires by the roadside, have adopted a telegraph-orientated way of life in this area.

## Chapter 6 - The Fauna

The birds line the wires like beads on a string for mile after mile. They are essentially birds of open country and, in their search for insects, frequently associate with other animals, even sitting on their backs as they feed, the birds watching for insects rising from the disturbed grasses

Cattle egrets *(Bubulcus ibis)* are abundant all around the lake and on the dune islands, associated both with the elephants and with all kinds of livestock. Like the carmine bee-eater, but much more frequently, they perch on the backs of the livestock, flying down at intervals to stalk along just ahead of or beside the grazing animal, capturing any insects disturbed in the grass. Flocks of them were to be seen at all the artesian overspill pools where the livestock were watered and in lagoons and inlets near villages and cattle pasture.

In the lagoons frequently used by the elephants, cattle egrets could sometimes be seen above the papyrus as they rode on the elephants' backs, or as they took off at intervals for brief flights when the elephants unseated them with their trunks. They, too, are insect eaters, being particularly partial to grasshoppers, locusts and crickets.

Another bird that was almost invariably to be seen near and on the lake was the West African black kite *(Milvus migrans)*. This bird is essentially a scavenger and, although it may swoop to catch small rodents on the ground, it is usually seen pecking at carcases, or among rubbish, or, especially on and around the lake, at fish refuse in and around fishing camps and villages. It sometimes also enjoys a feast of insects as when locusts or flying ants swarm, and then a number may be seen swooping and banking as they gorge themselves in the air.

As we drove towards the lake in August 1969, we found numbers of Abdim's storks *(Sphenorynchus abdimi)* nesting, the fledglings just leaving the nests, in and southwest of Maiduguri. However, we did not see any at that time near the lake. About mid-October, we saw some nests near Kukawa, and at the end of November we saw these storks congregating in large numbers around the rapidly drying pans near the Baga peninsula. *(See: Plate 8.5)*

Also in November, flocks, each of several hundred (and in one case over one thousand), of European white storks *(Ciconia ciconia)* rested for a few days on the Baga peninsula on their southerly flight. In December, wood ibises *(Ibis ibis)* were also to be seen in small scattered flocks; and at this time migrant waterfowl began to appear in the pans and archipelagoes of the lake and its vicinity. Of these, whistling teal *(Dendrocygna viduata)* was quite definitely the commonest species.

Chapter 6 - The Fauna

Diagram 14. The Sudan bustard, Ardeotis arabs sticheri. This large Bird is much sought after around Lake Chad by local sportsmen

One of the birds which is an integral part of the life of West Africa is the common or hooded vulture *(Necrosyrtes monachus).* It is to be found everywhere that man occurs, and is responsible for a great deal of the clearance of refuse in West African towns and villages, and, unsurprisingly, near and on Lake Chad, especially around fishing villages and waterholes where so many domestic animals finally lie down and die towards the end of the dry season and onset of the rains as a result of starvation. Wherever there is a carcase, there are the hooded vultures - that is, unless Rüpell's griffon vulture has also arrived on the scene. Rüpell's griffon *(Gyps rüpelli)* is an enormous bird, also a scavenger, which tolerates no rivals until and unless its own needs are satisfied first. These huge birds were very common on the western shores of the lake *(Plate 36.6)*.

Hawks and eagles are frequently to be seen everywhere on and around the lake. These include a very wide range of species, the ones that most particularly attracted our attention being the chanting goshawk *(Melierax metabates)*, of which several pairs held territories between Minetti and Portofino and, on the lake, the fish eagle *(Cuncuma vocifer)*. We did not see many members of this species, but isolated pairs occurred both along the fringe swamps and in the archipelagos of the lake. Of all African birds, this species is, to me, the one which represents the voice of the real 'Africa', for its wild ringing cry across the waters always speaks to me of vast distances, sunshine and freedom *(Plate 35.4)*.

In the savannah of the western and southern shores, and the inundation areas of the deltas, there are bustards: Denham's *(Neotis denhami)*, and the Sudan *(Ardeotis arabs)* 'greater' bustards *(Diag.14);* and the Senegal *(Eupodotis senegalensis)* and black-bellied *(Lissotis melanogaster)* 'lesser' bustards. All of these species are good eating and provide sport for considerable numbers of citizens from the larger cities in the vicinity of the lake. Indeed, at weekends, and on public holidays, it was not uncommon to see carloads of 'upper salary bracket' people of all races, armed with double-barrel shotguns,

## Chapter 6 - The Fauna

along the Maiduguri-Baga road in search of a fat bustard or two. Guinea fowl, once very common all around the lake, due to excessive hunting, are now no longer common, and those remaining are very wary indeed. We only encountered one species, the helmeted guinea fowl *(Numida meleagris) (Plate 36.5)*.

I was particularly interested to find out how the ostrich *(Struthio camelus)* population of the lake shores and hinterland was faring, because they also have been excessively exploited, especially in recent years. During the entire six months of the 1969-70 expedition, I saw (on several occasions) only one single flock comprising one cock and four hens which lived in the vicinity of Kukawa, and two other single hens. Only two eggs were reported to the District Head of Kukawa during this period, a fact which he bemoaned greatly, as traditionally ostrich eggs form an important delicacy in the diet of the Kanuri people, who also frequently use the shells to adorn the apices of their huts.

Although the District Head professed himself an ardent supporter of modern wildlife conservation, I had the feeling that this should - from his personal view-point relative to Kanuri country - undoubtedly be expected to include ostrich farming, which would provide him with his regular ostrich egg breakfast on a set number of days each year. It did not appear to occur to him, unaided, to admit that the Kanuri taste for ostrich eggs may account for the near-extinction of wild ostriches in Bornu.

Abyssinian rollers *(Coracias abyssinica)*, once a blue-bellied roller *(Coracias cyanogaster)*, Senegal hoopoes *(Upupa senegalensis)*, bulbuls *(Pycnonoyus spp)*, grey hornbills *(Lophoceros nasutus)*, turtle doves and mourning doves *(Streptopelia spp)*, speckled pigeons *(Columba guinea)*, the irrepressible weavers *(Ploceidae)* and an amazing variety of other small birds kept us company constantly with their calls and songs.

The speckled pigeons won a place of particular affection in our hearts with their extravagantly emotional courting gestures and cooings day after day on the roof of the house at Minetti. Another even built its nest aboard the *Jolly Hippo* as it lay at anchor for two weeks during December.

A little bird which was always around when we were on the lake proper was the West African little tern *(Sterna albifrons)*. These smart, friendly little birds were to be found on any little piece of vegetation, even on a single floating papyrus stem, on canoes and on floating islands. They were on the open waters and in the lagoons, or just winging with neat, quick strokes, low over the water. Storms did not seem to worry them: they just held on tightly to their floating perches, bobbing up and down on the waves as if they loved it. We had occasion, several times, to be grateful to these terns because, by perching on the almost invisible fish net floats, they warned us of

Chapter 6 - The Fauna

those submerged horrors that were ever ready to entangle our propeller *(Diag. 15)*.

Diagram 15. West african little tern (Sterna albifrons).

In the lagoons and archipelagos, several species of heron were found, perched on dead tree trunks or in the reedy shallows, along with white-breasted cormorants *(Phalacrocorax lucidus),* and even an occasional pelican *(Pelecanus onocrotalus).* It was fun, too, to watch the pied kingfisher *(Ceryle rudis)* hovering poised above the still waters, to drop like a fisherman's spear into the body of some unfortunate fish below.

Lake Chad is the home of the beautiful purple gallinule *(Porphyrio madagascarensis),* as well as of rails and moorhens: I found that the people of one village (immigrant Hausa people from Sokoto) were particularly given to keeping birds as pets, for here at one time they had a European white stork, a moorhen, a purple gallinule, and a crowned crane chick. West African crowned cranes *(Balearica pavonina)* used to be abundant in the northern parts of Nigeria, but excessive exploitation for sale alive to overseas zoos has reduced their numbers severely throughout Bornu, and possibly right across the northern part of Nigeria. I saw several small flocks in the inundation areas of the southern part of the lake shore in November 1969, together with large flocks of sacred ibises *(Threskiornis aethiopicus),* and one saddle-bill or jabiru *(Ephippiorhynchus senegalensis) (Plate 35.1 and 2).*

No account of the birds of Lake Chad would be complete without mention of that comical bird, so common around villages and refuse areas, the marabou stork *(Leptoptilos crumeniferus).* These birds were nesting at Gajiram, south of Minetti, in February, in untidy nests made of twigs. The local people also kept a number of these as pets around their houses, a particularly amusing one belonging to the local butcher at Kauwa Cross, who had taught it a few tricks. The bird was very popular, and the people gave the butcher pennies when it performed its tricks at his command. In flight this curious bird, with its tremendous wing span, looks far more graceful than the vultures as it rides the thermals high above the sizzling sand.

Of course, in mentioning selected avian species, I have omitted an enormous number of others: they were all such an integral part of the Chadian scene with its colours and its sounds that the shores and waters of the lake would feel barren indeed without them.

**Mammals**

In contrast to the bird life, the mammalian fauna of Lake Chad has never been rich in species, and is today poor not only in species but also in numbers[28]. Due to heavy persecution of the larger mammals, a persecution which continues to an ever-increasing degree almost unchecked even today, - these species are now represented by only a pathetic remnant. On the 1969-70 expedition I arrived with traps and firearms suitable for collecting a representative range of Chadian mammals. Indeed, I was even granted special unrestricted collecting licences for all species.

It was only when I actually began to investigate the true position, however, that I realised with dismay that every time I fired a gun it would serve as an immediate incentive to local hunters to follow suit even more avidly than they were doing already, and, secondly, it could only serve to hasten the decimation of species which, it had long been my hope, would one day form the nucleus of a Lake Chad National Park for wildlife on the Nigeria side. I did not therefore, shoot or trap any mammels except elephants.

I relaxed this decision only in the case of elephants, as I had been asked by the Forestry Department to investigate their size and commercial potential as meat for export to other parts of Nigeria.

A trapping programme for small mammals in the swamps, floating and dune islands, which was planned, also had to be cancelled due to a shortage of funds and, in consequence, personnel. This was the result, unhappily, of inflationary prices in Nigeria due to the civil war then in progress and in the Republic of Tchad due to two major rebellions which were disrupting normal services.

As regards the larger mammals of the Chadian shores, the descriptions of Denham in 1826,[29] and later of Alexander in 1907,[30] seem a far cry from today's scene:

'. . . the soil . . . clothed with grass and dotted with acacias and other trees of various species, amongst which grazed herds of antelopes . . . the lake . . . frequented by hippopotami . . . [and] . . . the pretty sight of a party of 5 or 6 white-rumped dorcas gazelles feeding among the bush . . . Their hindquarters and legs are pure white and make a sudden contrast to the rest of the body, which is an almost uniform chestnut, merging at a distance with the surroundings so that their forms are difficult to make out especially through the haze of the prevalent mirage.'

Nowadays, dama *(Gazella dama) (Diag.16)* and *dorcas (Gazella dorcas)* gazelles, spotted *(Crocuta crocuta)* and striped *(Hyaena hyaena)* hyaenas, and the cheetah *(Acinonyx jubatus)* are either extinct or almost extinct on the Nigerian shores of the lake. Dama and dorcas

*Diagram 16. The dama gazelle, Gazella Dama.*

*Diagram 17. The scimitar or white oryx, Oryx algazel.*

gazelles, however, still occur widely in the Republics of Niger and Tchad.

Red-fronted gazelles *(Gazella rufifrons),* bubal hartebeest *(Alcelaphus major),* and kob *(Adenota kob)* are still to be found around the lake in isolated, small groups. On the Cameroun side in the Chari and El Beïd delta areas and hinterland, a very noticeable and significant recovery of wildlife has resulted from stringent and progressive wildlife conservation measures, affecting many species but in particular the West African giraffe *(Giraffa camelopardalis),* the black rhinoceros *(Diceros bicornis)* and Derby's eland *(Taurotragus derbianus,* as well as the African elephant *(Loxodonta africana).*

*Diagram 18. Addax, Addax nasomaculatus, the fleetest of all the antelope to be found around Lake Chad.*

More recent conservation measures in the Republic of Tchad may be expected to result in a recovery of the scimitar or white oryx *(Oryx algazel) (Diag. 17)* and the addax *(Addax nasomaculatus) (Diag.18)* in Kanem. No comparable modern conservation measures have yet been undertaken in Nigeria and Niger, and it seems doubtful that the full range of species formerly found on their shores of the lake can now be recovered naturally. Three groups of red-fronted gazelles lived in the bush within a ten-mile radius of Minetti. One of these comprised a male, two females and a calf, while in each of the other two there was just one male and one female.

On various occasions also we saw red-fronted gazelles between Kauwa Cross* and Baga, but these were always shy and kept far away from the road. We also saw several between Gajiram and Malamfatori on the sand road. In September, the District Head of Kukawa presented me with a young female red-fronted gazelle which I called 'Twinkle'. She was already quite tame, having been hand-reared by the women in a local Kanuri compound. The District Head himself also had two tame sitatungas in his own walled compound at Kukawa. These are beautiful marsh antelopes still found in certain places in the lake swamps.

Twinkle was a very prim little gazelle, always reminding

---

* The small village and roundabout where the Maiduguri/Malamfatori and Kukawa/Baga roads cross each other.

me of the product of a 'best English girls' finishing school' of the 1930s. As more wild orphans joined the little collection at Minetti, Twinkle always remained at the head of the pecking order, assuming a delightful 'head girl prefect' attitude and taking pains to ensure that her senior status was clearly understood on arrival by each newcomer.

The second ungulate to arrive was 'Star'. She was a very young, very weak sitatunga calf. A hunter had found her hidden by her dam in the marshes and brought her first by canoe, and later by lorry, to Minetti. She was so weak and dehydrated that I was doubtful if she could possibly survive. It was difficult, even assisting her, to get her to swallow the warm, diluted milk that I prepared, but, with constant attention over the first few days, she began to pick up and became strong enough to share the improvised pen with Twinkle and two other orphans, 'Pumpkin' (a leopard cub) and 'Ginger' (a patas monkey).

The sitatunga *(Limnotragus spekei)* is a beautiful antelope specifically adapted for life in tropical African marshes. The hooves are elongate, and may be widely splayed as the animal walks over deep, soft mud. The habits of the sitatunga recall those of the North American moose, that huge semi-aquatic cervid, that loves to stand for hours submerged under water with only its head showing. The sitatunga also likes to stand in deep water among the reeds and papyrus, and does not hesitate to swim when necessary *(Diag.19)*. Its preferred food consists of marsh grasses and other soft herbs, and each pair of sitatungas occupies a well-defined territory. Usually only one calf is born at a time, but twins are not unknown. During the 1969-70 expedition I saw wild sitatungas on three occasions, and each time was absolutely enraptured with the beauty of this creature as seen in its natural habitat.

Standing, partially submerged in water among papyrus fronds and reeds adorned with flowering vines, while shafts of mottled sunlight filtered through, the dark sable, patterned coat of the male and the radiant chestnut coat of the female seemed more like mysterious shadows than living animals. Then the male stamps, with a slight sloshing in the water, lifts his head proudly so that his gently curving horns with their white tips come into focus, and stands motionless gazing intently at the intruder. And then, in a flash, the pair seem to glide and bound silently away to become lost among the sun-pierced fronds.

On another occasion, again while stalking elephants in papyrus, we found we were following a beautiful sorrel-coated cow sitatunga with her little calf at foot. The elephants had disturbed her and she was quietly following them down their trail. Then she heard us, stopped, sniffed and stamped her foot. She turned her head on its mobile neck, the great round ears spread wide, as did the calf also, a little image of its mother. And then quite gently, unafraid, they slipped away into the shadows of the papyrus. Not infrequently when canoeing in the lagoons in the late evening, we heard sitatungas barking to each other in the reeds, and once or twice a calf called with a plaintive nasal wail.

The pattern of chevrons, stripes and spots on the rather shaggy water-resistant coat is entirely individual and, as viewed from above is asymmetrical on the two sides in the one animal. Sometimes, stripes and patches are almost entirely lacking. Only the male has horns of which the record length in West Africa is 88cms. He attains a height of about 120cms at the withers or shoulder, and may weigh about 200kgs. He may have a white mane along the neck and spine.

Diagram 19. A pair of sitatungas: I was absolutely enraptured with their beauty when seen in their natural habitat.

Sitatunga hides are very popular as reclining mats in Kanuri homes, as they are said to enhance fertility in those who sit on them. They are extremely similar in colour and pattern to hides of the bushbuck or harnessed antelope *(Tragelaphus scriptus),* common near rivers in savannah country. The bushbuck, however does not share the sitatunga's semi-aquatic habits, or its specially adapted hooves. Possibly, also, the hair of the sitatunga is a little more shaggy, and the hair texture (unless spoiled during tanning) feels softer and more separated than that of the bushbuck. When the animal's hide is tanned, it can be extremely difficult to be sure of the species unless the animal was a very large one, as the sitatunga may grow considerably bigger and taller than the bushbuck.

In the past taxonomists have tried to distinguish different races on the basis of skin pattern in both the sitatunga and the bushbuck, but it is quite evident, when one sees the variation in individual pattern in hides collected in the same locality, that there is no sound basis for this type of classification.

When Star joined Twinkle she was, even then, at only about a week or ten days old, taller and longer than Twinkle and much more gangly in appearance due to her long, flexed limbs, elongate hooves and head position. The head is generally held low and forward. She took long, low steps, or long bounds that looked as though they were performed in slow motion. Twinkle, on the other hand, adapted for life in the hard, dry savannahs, took quick, short, neat steps, flicking her tail constantly as if nervous. Her little hooves were dainty and trim, and she held her head high - a little haughtily, perhaps.

Her eyes were small and sharp, her ears smaller in proportion to her head, and her nose a bit 'roman' in style; whereas Star raised her head only when she stopped and stood still. Then she would spread her great round ears and look out through large, limpid, shade-adapted eyes with a gentle, self-effacing look. Star was definitely more near-sighted than Twinkle, the gazelle of the open grasslands and savannahs. Then again, Twinkle's coat was short and very trim and neat, with its longitudinal black stripe, whereas Star's coat was long and a little shaggy, as if perhaps it needed brushing. Standing together in the pen, these two seemed to epitomise the perfection of their adaptation to their two completely contrasted environments existing within the same neighbourhood.

In late November another sitatunga orphan joined our menagerie: this was called 'Minetti' after the name of our base camp. She was quite distinct from Star as she had a sorrel coat, whereas Star's was basically a light chestnut colour. Although younger when brought in, she was in much better condition and throve from the beginning. Both these sitatungas later went to the Jos Zoo (Nigeria) to join 'Yobe', a much older male with a sable-coloured, pattern-free coat, caught near Malamfatori *(Plate: 8.6 and 7)*.

The dorcas, or white-rumped gazelle *(Gazella dorcas)*, is generally smaller in size than the red-fronted gazelle, and typically lacks the longitudinal, black side stripe. Like the Thomson's *(Gazella thomsoni)* and Grant's *(G.granti)* of East Africa, the range of the red-fronted and the dorcas overlap and, moreover, intermediate forms occur. As in the case of Thomson's and Grant's gazelles, the occurrence of intermediate forms within mixed herds is suggestive of interbreeding, although convincing evidence to prove it is apparently still lacking.

Around the northern and north-eastern shores of Lake Chad, both the slender dama *(Gazella dama)* and the larger, white, or scimitar oryx antelope *(Oryx algazel)* occur. The former stands some 90-100cms at the shoulder, with sturdy, annulated horns some 25-30cms long; while the latter stands 120cms at the shoulder and carries slightly curved, slender, sabre-like horns which may reach 115cms length. The colouration of the dama is red-russet on the back, flanks and part of the head, the rump, belly and part of the legs being white. The tip of the tail is black. The colour of the oryx is white with chestnut, brown or greyish markings on the face, neck, lower flanks and upper parts of the legs. The tail tip bears a tuft of long, dark hairs. It is interesting to note that these desert ungulates so often have a whitish general appearance and dark tail tip, whereas the underside of the tail in darker coloured, shade-dwelling, tail-raising types such as the bushbuck and sitatunga is frequently white. The tail appears to be a communicatory organ in many ungulates, especially between parents and their young.

Both the dama and the oryx migrate seasonally along routes with a north-south alignment, moving south in the drier weather and north in the rains, but, whereas the oryx prefers to live in small herds (a habit disrupted under heavy hunting pressure) not exceeding twenty animals, the dama gazelles may gather in herds of up to one or two hundred at certain seasons, while at other times they revert to groups of not more than two or three animals.

Antelope and gazelles typical of the Sudan and transitional Sudan-Sahel savannahs tend to occur more on the Cameroun side of the lake. They include a variety of species including the bulbal hartebeest, duikers, waterbuck and also, further south, giraffe, black rhinoceros, Derby's eland and elephant, for example, together with leopard, smaller cats, hyaena and even lion in the adjacent hinterland.

The common jackal *(Canis aureus)* is not infrequent around most of the lake shores, and the wild hunting dog *(Lycaeon pictus)* is said still to occur in the Chari and Logone basins. The pale fox *(Vulpes pallida)* also occurs widely in the vicinity of the lake, and the fennec *(Fennecus zerda)*, an insectivorous and frugivorous fox-like animal, is said to occur to the north and north-east of the lake.

We frequently encountered civets *(Civettictis civetta)*, white-tailed mongooses *(Ichneumia albicauda)*, the wild African cat *(Felis lybica)* and the Lake Chad hare *(Lepus chadensis)* on the road at night between Portofino and Minetti. We also collected one male honey badger or ratel *(Mellivora capensis)* which had been killed by a vehicle, and one zorilla *(Ictonyx striatus)* on the road in Maiduguri township one night. Then there were aardvarks, or antbears *(Orycteropus afer)* which dug huge burrows around pools and pans and in the open savannah, and a variety of rodents, including the jerboa *(Jaculus jaculus)*, gerbils *(Taterillus lacustris* and *T.gracilis)*, a variety of rats, mice and shrews, ground squirrels *(Xerus erythropus)* and crested porcupines *(Hystrix cristata)*. We noted an abundance of bats living in the floating vegetation both on the lake and on shore, but we did not make any attempt to study these systematically.

On one occasion, Mohammedu and I saw a black ground squirrel cross the road between Maiduguri and Gambaru (to the south-east of the lake); and on three occasions we saw an albino ground squirrel which apparently lived just near the milestone seventy miles from Maiduguri on the way to Baga. Unfortunately we were never able to get a picture or trap it for examination. In January I acquired two crested porcupines which I called 'Prick' and 'Prod', and these provided us with a good deal of hilarious entertainment.

Then there was 'Blob', a minute dormouse orphan which I found under a tent which I had stored at Maiduguri in August. When I found him, he was completely pink and naked but we successfully hand-reared – or should I say 'tongue reared'? – him on mixed condensed milk and saliva which he licked direct from the tips of our tongues. He grew a very lovely, soft grey coat and fluffy tail, lived eight months, and died, apparently from pneumonia, during the rainy season. We believed this dormouse to be Buchanan's dormouse *(Graphiurus olga).*

Hedgehogs *(Atelerix spiculus)* occur around the southern shores of the lake, but we found no evidence of them further north than Gajiram. The African hedgehog is much smaller than the European species and very easy to tame. It seems to like termites mixed with minced meat in captivity, as well as milk and egg.

The most delightful of all our Chadian orphans was 'Pumpkin', later to be renamed 'Sultan'. Pumpkin was a leopard cub *(Panthera pardus),* one of a pair of twins captured somewhere up the El Beïd river on the Cameroun border. The hunter who had shot the mother sold the cubs to the Inspector-General of the Nigeria Police, whose home was in Maiduguri, and who had left them in the care of his household. The cubs, however, did not thrive and one died. The other was taken to the Veterinary Department but, as they had no suitable Government facilities for this kind of patient, someone suggested asking me to try to care for it.

## Chapter 6 - The Fauna

Not knowing anything about all this, I was a little anxious when I was stopped by a constable on a routine trip into Maiduguri on 29 August 1969 and told to proceed to Police Headquarters immediately.

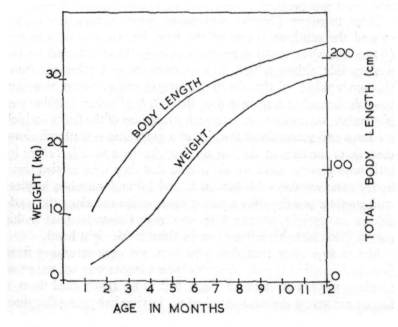

*Diagram 20. Graph showing the growth (body weight and total length) of the leopard cub, 'Pumpkin', during the first year of his life.*

With some trepidation I duly complied, and was ushered into the office of the Assistant Commissioner who said: 'We have got a leopard cub which belongs to the Inspector-General but which is very sick. Please will you treat it for us?

So, instead of finding myself arrested, arraigned and interrogated, I found myself being swept away to the Veterinary Department where the funny, dirty, weak, pot-bellied little scrap was handed over to me. So pot-bellied was the little animal that I promptly named it 'Pumpkin', pumpkins being the only vegetables that we ever found then under cultivation on the lake islands.

We hastened back to Minetti where I treated the cub first with Epsom salts, then sulphur drugs and antibiotics, leading on to a diet of shredded liver and warm milk laced with brandy, glucose and eggs. Within two days it was much better, and then I washed it in a bubble bath containing disinfectant. With all this handling and care, Pumpkin quickly became very affectionate and responsive, and it was soon possible to put him in the pen with Twinkle and Star. On arrival at Minetti he weighed 3kg and was 62cms long from nose to tail.

Patas monkeys *(Erythrocebus patas)* were extremely common around the southern shores of the lake, but we saw no baboons *(Papio anubis)* or tantalus monkeys *(Cercopithecus aethiops)* on the western side, although both these occur to the east. When the fruits (Sodom apples) of the shrub *Calotropis procera* were ripening, towards the end of August and on through September, October and November, we noticed ever-increasing numbers of the fruits - which are large and green about the size of a grapefruit - scattered along the side of the tarmac.

## Chapter 6 - The Fauna

At first we thought these were left there by little boys playing, until we discovered that they were carried there by the patas monkeys. Mohammedu and I found ourselves helpless with laughter one day over a young patas which ran along the roadside in an upright posture with one huge *Calotropis* fruit in his mouth, one under his left arm and a third in his right hand.

Eleven days after Pumpkin had joined us, Mohammedu caught a very young female patas monkey which we named 'Ginger'. She was to become Pumpkin's bosom pal for the whole of the first year of his life. Indeed, I believe that it was largely due to Ginger's influence and constant closeness to Pumpkin that he developed towards maturity as an incredibly gentle leopard. Ginger never really made friends with any particular person, but was wholly absorbed with Pumpkin from the moment they were introduced and accepted each other as babies. When Ginger arrived she was the same length as Pumpkin, with the same proportion of tail to body, similar agility and a somewhat similar colour. Their mouths were about the same size, and in each case their eyes both looked straight forward. It was apparent that they imagined they belonged to the same species, and Ginger even went so far as to adopt many of Pumpkin's feeding habits.

Both grew together in agility at climbing, running, romping and tumbling. After about a week following Ginger's arrival, I shot a pigeon for Pumpkin. At first he did not understand what it was or what it was for. Then I cut open its abdomen and smeared some of the contents with some of its blood on its feathers and teased him with it. He soon got the message, took it and began to play with it. Then he got the taste and tried to eat it, but his paws and his claws got out of control: then the feathers stuck to his nose and got in his eyes and teeth. Ginger was fascinated and licked a piece of the exposed raw flesh. Pumpkin made a complete mess of his first pigeon but, with more experience, he soon mastered even the feathers, while Ginger acquired a taste for raw meat and used to sit beside him, a torn-off leg or wing in her hands, pulling away the feathers and chewing up both flesh and bones - a carnivorous monkey *(Plate 9.5)*.

The pen was now occupied by Twinkle, Star, Pumpkin and Ginger, with Twinkle assuming the dominant role. Pumpkin and Ginger would romp together by the hour, Ginger climbing up into the roof if she got tired or found Pumpkin too rough. Ginger, however, never hesitated to bite him in return if he nipped her. A favourite game was to lie together on the roof of a little shelter in one corner of the pen when Twinkle and Star were sleeping inside.

Pumpkin would lean over and try to annoy them with one paw. Sometimes he succeeded in making them come out, and then he would pounce on to their backs only to tumble off over the other side as they leapt away. But at other times they were too sleepy to respond. Then he would lean further and further over the edge in his efforts to touch them with his provocative paw, until finally he would lose his balance and tumble off on to his head in the sand.

Not only did he, in his play, demonstrate and develop the instinctive tendency for a leopard to spring down on to its prey from above, but he also knew how he ought to leap up from the ground up on to the shoulder of his prey, pulling the neck down smartly with one paw to break it. He tried this latter technique very clumsily on both Twinkle and Star over a period of several weeks, but of course he was not nearly strong enough yet to hurt them. Star was tolerant of this game, but tired quickly. Twinkle seemed to feel that the whole idea was undignified and, after a mere couple of attempts by Pumpkin to spring on her back and break her neck, she would quite calmly, but defiantly, turn round and butt him hard on the head. This usually chastened him for a while, but he soon forgot and tormented her again *(Plate 9.4)*.

Pumpkin soon lost his pot-bellied appearance, and developed a sleek yellow coat with dark spots and rosettes, and a superb white undercoat to his abdomen and immensely long tail. As we were now spending more and more time on the lake, and the orphans were becoming bigger and stronger, they were handed over into the care of our good friends, the Barlows, who lived near

Jos, until they should be sufficiently mature to move into the small zoo at Jos. Ginger and Pumpkin remained together until Pumpkin was a year old and weighed 31kg and was 212cms long *(Plate 9.6)*. Throughout this time he was handled daily and taken out for walks and tree climbing on a lead. He never put a claw wrong and remained completely gentle.

In August 1970, he was moved into a new large enclosure at the zoo, together with 'Candy', a nine-year old female leopard. Unfortunately, Ginger and Candy were initially incompatible, and Ginger was injured by Candy and had to be destroyed. It was several months before Pumpkin (now called 'Sultan') and Candy became fully compatible, but, at the time of publication, they are on completely intimate terms with each other. What is more, I could still safely put my hand between Pumpkin's magnificent jaws.

The hippopotami *(Hippopotamus amphibius)* of Lake Chad have already been mentioned and, although they seemed to be widespread on the lake, it is clear that they are nowhere really abundant in it today. They are more abundant in the Chari River delta than in the lake proper. Under effective conservation management, the lake could certainly support a much greater hippo population than at present. Unfortunately, gillnet fishing and the presence of hippopotami are incompatible, and it would be necessary to define suitable protected areas for the hippos from which such fishermen would be excluded. It is quite common in Lake Chad for hippos, intentionally or inadvertently, to overturn canoes, sometimes biting the canoe in two and mauling the occupants. In many cases the occupants are drowned, although most Yedina fishermen are good swimmers and, unless injured, would escape. Moreover, the hippos sometimes become entangled in gillnets or foul-hook lines and become exhausted, drowning or being speared to death by the fishermen.

The hippos generally come ashore at night to graze, neatly clipping the grass short, leaving a double track wherever their regular paths lie. In contrast, the elephant always leaves a single track, as it places its feet in front of each other on the ground so that the right and left footprints overlie one another. Where hippos live entirely among the floating islands, it seems that they must eat the vegetation while standing within the shallows as they can never come ashore.

It was natural for me, as a specialist in elephant studies, to pay very special attention to the Lake Chad elephants. These are the resident elephants of the actual lake, and are to he distinguished by their behaviour pattern from the migrant elephants which ply between the lake and the far hinterland in the Chari/Logone basin and southwards through Gwoza to the Gongola River. Both groups, however, belong to the same variety, *Loxodonta africana*

*africana* (the african savannah elephant). Their use of the Gwoza route is in some doubt nowadays (1969/70), due to ever-encroaching cultivation, but the elephants of Cameroun have increased enormously in numbers since the introduction of effective conservation measures. The Lake Chad elephants have never been protected, and are available to hunters in Nigeria on payment of a licence fee of N£10 for three elephants,* plus a graded trophy tax for each of the three sets of tusks obtained, of N£10, N£20 and N£35 respectively. The result of this has been an inevitable dwindling in numbers of the Lake Chad elephants to a total of less than one hundred.

I had been asked to look into the potential commercial value of these elephants in terms of meat marketing, so it was necessary for me to collect some for measurement and examination. I also wanted to find out if it was true that these elephants had outsize feet, as was widely believed among expatriates with an interest in hunting.

The Kanuri themselves do not kill or eat elephants as they believe the elephant (Kanuri: *kuman,* pronounced 'Koomarn') to have been the ancestor of man. The elephant is to them much more readily accepted as a relative than any of the sub-human primates!

In 1962, when hunting Lake Chad elephants near Ngurno I had seen the biggest elephant I have ever set eyes on, either during my former elephant hunting

---

* 1970

activities or subsequently, during my research on elephant diseases in East Africa. I wanted to know if this one had been a freak or whether the Lake Chad elephants, like the Lake Kanuri, the Yedina and the ancient So people, grew unusually tall. On that occasion, while manoeuvering to shoot the giant, a smaller bull saw and attacked us, and I was obliged to kill him with an emergency shot at 10m range as he charged!

In consequence, I lost the giant, which fled with the rest of the herd at the sound of the shot. It was not that he carried exceptional tusks, but that he stood at least 30cms higher at the shoulder than any of his companions in that particular mixed herd - a herd that numbered over 100 animals, including cows and calves.

In November 1969 Mohammedu and I began hunting elephants in the lagoons along the Nigerian shores. Hunting elephants in the swamps of Lake Chad is, to put it mildly, tough going. Already having a number of successful elephant hunts in savannah forest and montane conditions behind me, I did not at first realise that hunting the Lake Chad elephants in the lake would prove to be an exercise of an infinitely more demanding order than anything I had ever experienced in this way previously *(See Plate 6.3)*.

Initially we made a number of exploratory trips and discovered that, at that season, the main body of the elephants were living as a single mixed herd of about seventy cows, young bulls and calves with two attendant sire bulls which were said to be very large. These preferred the lagoons of the shore line from Ngurno to Dowasheri. In this area there was said to be another bull group of three animals and a loner which lived on a particular island off-shore from Ngurno. In the lagoons on the northern side of the Baga peninsula a herd of eight bulls was living.

These were the ones seen on our first trip in the *Jolly Hippo*. In this herd, there was one notorious elephant nicknamed by the local people after the self-styled leader of the secessionist rebel forces of so-called 'Biafra' in the then current Nigeria civil war. This elephant lacked the tip of its tail, was very large and mean, and had already killed several people, so I had been asked to try to kill it.

On our first armed encounter, we located the bull herd in the lagoons about five miles from Portofino. We knew where they were, long before we could see or hear them, by the flock of egrets that accompanied them, and which every now and then took off and flew around a bit, before settling on their backs again. We found them about 10.00a.m. quietly standing in a lagoon, some feeding on the grasses and others lackadaisically spraying saturated mud on their bodies. We edged along in the canoe, feeling very exposed and vulnerable, until we were within 20m of the nearest.

## Chapter 6 - The Fauna

There was a dead tree stump standing in the middle of the lagoon, so I climbed out of the canoe into the saturated black mud which was about a metre deep at that place and sighted the rifle at his temple, using the tree stump to steady the rifle. However, the wind-induced current was irregular and very strong both against my legs and also against the tree stump. Moreover the canoe kept bumping against me, so any attempt to shoot would have been suicidal. I carefully, slowly and silently climbed back into the canoe, but not without unintentionally alerting one elephant which gave the alarm, and they all moved off across the lagoon and into the papyrus fringing it.

On the next occasion, we found them in the same area, and this time we followed them up the bank into the papyrus fringe. Emerging from the black saturated mud, which filled our clothes and shoes, on to the trembling peat bog, did not enhance our confidence. I was within 30m of my selected animal - an enormous elephant, - but the peat was unstable and I felt myself sinking. Moreover there was a stiff breeze, which, while it contributed to our stalking success, was too strong for a certain shot, without some object for stabilising the rifle. It was all very tricky and unpleasant.

Soon after this episode, during the visit of Mrs Barlow and her brother Mr Frank Drew, we went into the lagoons again in the hope of finding elephants, as neither of our visitors had ever seen wild ones close up before.

The elephants were in a convenient position and we took photographs. Then they moved off and we manoeuvred the *Sportyaks* into another lagoon from which we might be able to go ashore and get a clearer view.

We followed the elephants over the grassland for about 1.5km, and then entered boggy papyrus. Here we were able to approach to within about 50m of the herd, obtaining a good view, hearing distinctly the slap, slap of their ears against their sides, and their pop-pop-popping stomach rumbles. I had the rifle in case of emergencies, but we took no risks and withdrew again in good order to the boats. Then we again got into a new position to watch them where they had returned to the water's edge. The breeze now carried our scent towards them and they were uneasy, one of them walking about irritably as if hunting for us.

Our friends wanted to smoke, and it was most amusing to see how, the minute they lit their cigarettes and the scent wafted towards the elephants, each raised its trunk, the tip bent towards us as they identified and located the source of the smoke. One of these elephants had a broken ear which apparently worried him for he kept touching it with his trunk, and then dumping tangles of vines and green leaves on top of his head. The great, mean one was there, as was one tremendous animal that I earmarked for future study and which was in fact the second one that I collected subsequently.

Chapter 6 - The Fauna

Our friends were delighted with this experience which certainly contained an element of excitement and contact denied to the usual 'national park' type of game viewing for tourists.

About a week later Mohammedu and I decided to try again to get near enough to shoot the mean, dangerous elephant in this herd. The weather was uncertain, but we started out early with clear sunlit weather, using both *Sportyaks.* We duly located our elephants in the lagoons but they were in a rather tricky position for an approach, wallowing and rolling in the mud downwind from the approach route across the water.

With great care we paddled the boats across, and then Mohammedu and I climbed out into the saturated mud. This proved too thick and sticky for safety and we attempted to climb aboard again. A slight diversion was created when my foot skidded on the wet boat and I fell back into the mud. This upset the elephants which walked away across our bows in a very determined manner through the water. They looked superb. Saturated and muddy, we set off again across the lagoon, and leaving one dinghy in the lagoon and one in the papyrus, we squelched up their path on to the grassland. A local fisherman accompanied us as we did not want to lose our way.

After about 1.5km of terribly hot, dry walking, during which the wet mud dried into our clothes like black starch, we caught up with the herd where they had entered a papyrus bog based on trembling peat. Clouds had been gathering during the past half hour, and now looked angry and stormy. A strong wind began to blow fiercely from the east, heralding a storm. After our earlier attempts to get a safe shot at an elephant in this marshy area, I had decided to use the technique so greatly favoured by that famous hunter of African elephants, Karamojo Bell. He used a surveyor's tripod with a sandbag on top, on which to rest his rifle. So now I had brought my heavy-duty cine-camera tripod complete with sandbag[31].

We set up the tripod and sandbag behind a thin clump of papyrus as close as was reasonable to the nearest elephant, the mean one, then about 35m from us, and waited. The wind grew gusty and dangerous, then turned into a cold gale bringing stinging rain. We stood there, drenched to the skin, teeth chattering, shivering from head to foot. The elephants moved into a bunch and huddled together for a while. Then, apparently without any reason whatsoever, the one with the broken ear started walking straight at us. It was quite unaware of our presence, but in that bitterly cold driving rain, the bog as slippery as ice under our feet, I knew there was no retreat. I lined up the rifle resting it on the soaking sandbag, in case he attacked, saw it trembling like a leaf in my shivering, cold hands.

The elephant circled round us at a distance of about 100m, the others following him in single file. Then they all set off on a 10km walk across the grassy headland, doubling back eventually towards the lagoons. We followed, half running, half stumbling, first through the boggy tangle of vines and papyrus with its huge wet potholes made by the elephants' feet and then across the grassland with its hard fallen grass cover of the broken stems of dry grasses and reeds all about 30cms in depth. Then the clouds dispersed and the sun shone again with scorching heat, so that we found we were gasping with thirst and sweating like horses.

Once again we caught up with the elephants where they were drinking at a pool in a swampy *Vossia* grass meadow. We stalked on hands and knees through the shallow water until we were about forty metres from them but we could not go closer. We knelt quietly in the water. The peat here was as insecure as in the first swamp. They began milling around angrily, coming alarmingly close to us as we crouched in the water and mud. I felt a nibbling at my ankles and saw the pool was full of *Polypterus* fishes, swimming in rings around us. It was all very disconcerting, This time I knew there could be no further reliance on the tripod set in the quivering bog, and took two quick free-standing shots at the mean one, one into his temple and one into the heart. The other elephants felt things were getting too dangerous and hastened away.

It was 6kms walk back across the scorching grassland to where the boats were left. Mohammedu found some wild 'cape gooseberries' which were refreshing to eat and certainly helped us all to endure the last lap of that hair-raising hunt.

The following day was spent obtaining measurements, extracting the tusks, and making observations on the carcase, while the local butchers quartered it. The butchers turned out to be quite remarkably inefficient at this, and by the early afternoon my patience wore very thin when they said for the umpteenth time that it was 'quite impossible' to separate the head from the body for tusk removal, and we 'must wait until they had removed all the meat from the carcase'.

Not being exactly an amateur at this business, and realising that they really wanted me to go away so that they could get the tusks illegally for their own profit, I organised our own small group to haul on the head while I cut it clean off, by muscle-by-muscle dissection, and careful inter-vertebral separation, using only a surgeon's scalpel with No. 24 detachable blades. The butchers were sufficiently impressed by this demonstration actually to lend a hand then with the tusk extraction!

Mohammedu and I spent many more hours after this hunt, tracking the various groups of elephants across grassland, and by canoe through the swamps. We never obtained an absolutely clear view of the whole mixed herd, though we saw them often in papyrus, the trunks, heads and backs of the adults showing, but the calves mostly remaining completely hidden.

We killed two more bull elephants, including the huge bull in the Portofino herd. He was indeed unusually large, for he measured 3.6m at the shoulder (straight measurement not over contours) and his forefoot circumference was 180cm. Both of these measurements exceed the maxima recorded to the present time (1970) anywhere in Africa except Angola. His tusks weighed 27kg each side. His relative molar age was FM.V/10 - about 45 years. He had thus by no means attained his maximum size or weight, nor his maxumum tusk size. His heart/body weight index of 0.5g heart weight/100g body weight, indicated a body weight of 6,200kg (just over 6 British tons)[32].

This great bull elephant carried on his flank and head old scars from the efforts of former hunters to kill him. He was the one mentioned earlier with the abnormally enlarged anal hernia, the cause of which we could not determine. He also had an abnormal extra molar behind the 6th ( normally the last) in the right jaw.

I also decided to kill the loner, because it became evident to me from the description by the villagers of his habits that he must either be injured or sick. I found that he had indeed been injured and was very sick, although still feeding adequately to keep going. A large iron, poison-laden arrowhead, with its tail bent as if it had been fired from a muzzle-loading gun, had passed right through the pneumatic sinuses of the skull, missing the brain, and had entered the muscles of the neck. Here it had lodged itself, leaving a foul, infected fistula, crawling with maggots, right through the skull to the arrowhead. There was no pus around the arrowhead but in the vicinity of the poison there was a mass of green degenerating flesh. Septicaemia had become generalised, even the inner surface of the tusks, where the ivory is formed from the cone of inner pulp, being discoloured and diseased.

These elephants had indeed provided me with some interesting data, and it was clear - they were all very large compared with East African bull elephants of equivalent age - that the Lake Chad resident elephants do grow unusually large and tall. However, their feet are in no way abnormally large compared with body size, although the footprints in sand or mud may *appear* larger than those of average sized elephants elsewhere. It was also clear that their tusk growth is in no way abnormal, and that if allowed by hunters to attain maximal age, the Lake Chad elephants would then carry very heavy tusks indeed, as elephant tusks continue to grow throughout the elephant's life. As the elephants I killed included the one that I judged to be the oldest living one in the area, and that turned out to be only middle-aged, it seems that hunters have already managed to eliminate all the really old giants that formerly lived in this area, and those must indeed have been enormous animals. I have frequently wondered if the tendency of elephants and people to grow unusually tall on Lake Chad is due to some dietary factor - perhaps the presence of the trace element selenium in the water. Selenium, for example, is known in certain conditions to enhance the longitudinal growth of limb bones.

While we were at the lake a total of nine elephants (including my three) were legally killed, and at least seven others (including at least one cow elephant) known to have been wounded and lost by the hunters who killed the other six. The arrowhead lodged in the head of the loner that I shot was tangible evidence of former illegal hunting with an illicit weapon, while the ancient scars and injuries seen on the first two I collected, and on the other elephants we observed, testified to the degree of persecution these elephants suffer.

The evidence available to me is fully adequate to state, in no uncertain terms, that unless a halt is called both to the licensed and to the illegal hunting of elephants in this area, some of Africa's most magnificent elephants in one of the most ancient habitats frequented by the species, will very soon be lost to the world. The unrestricted killing and injuring of the few remaining healthy sire bulls has already reached a point which has seriously reduced the breeding potential of this isolated, dwindling and important little population of elephants.

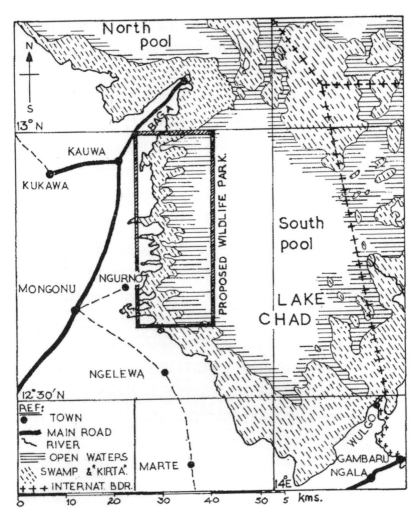

Map 12. The south-western shores of Lake Chad showing an area which would be suitable for development as a wildlife park

What is really needed is the total protection of an area not less than 15km wide running from north to south from Baga to a point just south of Ngurno such that it extends from the savannah on the west to the eastern edge of the floating vegetation which fringes the western shore. This area, shown on *Map 12* comprises about 600kms$^2$. Then after a period of total protection to ensure breeding success and the full recovery of population numbers - that is after about fifteen years - there would be an endemic elephant population of about 500 Individuals. Additional immigrant elephants would, however, radically alter this estimate. In any case a scientifically planned and operated annual harvesting programme could then be opened.

Such an area,* designated and managed as a national park, would of course also provide for habitat protection and management in favour of all the wildlife, and particularly the protection of the beautiful sitatunga, otters, hippopotami, crocodiles and other interesting reptiles, as well as a sample of the Chadian avifauna. It could in due course be opened for tourism providing boating (especially canoeing, kadai-ing and sailing), wildlife viewing, bathing and photography. Genuine miniature papyrus *kadais* could be locally made and sold at prices from N£2 to N£10 according to size, as they are already to the privileged few.

A recreational area of this type could not fail to attract revenue from tourism, providing that attractive hotel facilities and readily available travel visas were also ensured.

---

* See Chapter 12

Conservationists (myself among them) have repeatedly recommended during the past fifteen years the establishment of a wildlife conservation reserve or park in this area, but I wonder if the Nigerian Government will succeed in making up its mind to protect this potential paradise before it is too late.

*What an interesting and profitable recreational wildlife park it could be. . . .*

# PART III

# THE PEOPLE OF LAKE CHAD

# Chapter 7.

# Conquest, bloodshed and tyranny

The known history of the people of Lake Chad is inextricably bound up with the rise of the Kanem-Bornu Empire[33][34][35][36][37]. The story begins in the eighth century A.D., when groups of non-Muslim Zaghawa Berbers, Arabs of Yemenite origin, and possibly some Jews as well, moved into the southern parts of the Sahara desert west of Nubia. There is evidence suggesting that these migrations were not primarily invasions of a military nature, but rather represented emigratory withdrawals in the face of the fanatical advances of the then young religion of Islam *(Map 13)*.

The most powerful of these groups, the Beni-Sef, under their leader, Sultan (or Mai) Dougou, had, by the beginning of the ninth century A.D., already established the Sefuwa (or Dougou) Dynasty - a dynasty that was to last almost one thousand years with its capital at Birnin Jime (City of Jime, or Njime) in the land of Kanem, north-east of present-day Lake Chad.

The Sultan, *Mai* or king, was the authoritative head of the kingdom, and was supported by a Court of twelve councillors. Members of the royal family were appointed to

## Chapter 7 - Conquest, bloodshed and tyranny

senior administrative posts in outlying districts. The class system in operation was, as may be expected, based on a system of slavery, the wealth of Kanem depending largely upon the export of slaves and ivory.

Sultan Hume (pronounced 'Hoomeh') (1085-97 A.D.) was the first Sefuwa king to be converted to Islam, and it was under his Muslim successors (the names of the most outstanding of whom are known from written records)[38] that Kanem's power was greatly extended both westwards and southwards across the Chad basin to embrace the whole of Bornu, reaching its apogee as a mighty empire in the fifteenth century A.D. It is recorded that Sultan Dougou's original capital was a tented city, in the environs of which grazed horses, cattle, camels, sheep, goats and donkeys, and where the people cultivated beans and millet. Although the early Sefuwa immigrants to Kanem were resistant to the influence of Islam, this resistance did not last long, and after Sultan Hume's conversion, the culture of Islam, together with the skills of reading and writing, soon dominated the way of life of the rulers of Kanem.

In those days the land of Kanem was apparently much wetter than nowadays and the oases and pastures more extensive. It is fascinating to allow one's imagination to try to picture the scene, probably little different from the way of life as one sees it today along the northern and western shores of Lake Chad. This outstanding ruler was not only a great administrator and military commander, but also a devout Muslim. He endeavoured to establish Muslim law throughout Bornu, building mosques in brick, and even a hostel for Bornu pilgrims visiting Mecca. Unlike his predecessors, however, he took the precaution of minimising the dangers of intrigue, by retaining the members of the royal family at court, endowing them with vain titles of apparent rank, while filling the administrative outposts of his empire with men of humble (even, in some cases, slave) origins.

As today (1969/70), so then, one would see the camel drovers with their wares for trade and commerce, armed with swords, daggers, spears and bows, their faces swathed in the dark blue cloth dyed with locally-grown indigo, or hand-woven unbleached linen, to protect them from the blowing sand and dust, wrapped about with voluminous, but cool, robes, and shod either with tough raw-hide sandals or else soft, supple boots of camel leather. One can imagine the herdsmen pasturing their flocks of sheep and goats, 'white, ring-straked and speckled'* under the burning sun around the mud pans, seeking out each tiny patch of shade permitted by the stunted bushes. One might see the horsemen on their gaily clad stallions - some tough wiry ponies from Barbary; other 'aristocrats' from the Bahr el Ghazal - hastening across the sand dunes; or maybe a lonely leper devoid of nose, hands and feet, riding a thin little donkey, patiently

---

* The Holy Bible KJV Genesis. 30 v 39

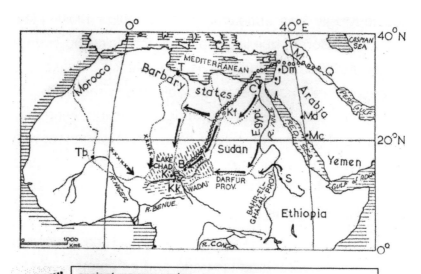

..............  ancient caravan routes

──▶  possible immigration routes used by the Zaghawa, Berbers and Yemenite Arabs via Dafur, and by Jewish refugees from the Middle East and Barbary states.

xxx▶  main approach routes of the Fulanai conquerors of the Hausa states west of Kanem and Bornu.

o o o o o  a link between the ancient Yedina of Lake Chad and the Arabs of the Qurna marshes has been suggested. Local tradition suggests that the Yedina may have come to Lake Chad in the time of Noah's Flood via the Bahr el Ghazal and Kufra Oasis from Mesopotamia. Perhaps such a route was feasible during one of the transgressions of Lake Chad or Mega-Chad.

|||  Kanem        ≈≈≈  Bornu

| B | supposed site of Birninjime | Kk | Kukawa |
|---|---|---|---|
| C | Cairo | M | Mesopotamia |
| Dm | Damascus | Mc | Mecca |
| J | Jerusalem | Md | Medina |
| K | Kasr Kumo (Fort Calabash) | Q | Qurna marshes |
|  |  | S | Sennar |
| Kf | Kufra Oasis | T | Tripoli |
|  |  | Tb | Timbuktu |

Map 13. The origins of the Sefuwa dynasties of Kanem an Bornu.

and longingly begging alms from the Faithful of Islam. (See *Plate 32.5*)

As time went on the ruling dynasty integrated with the local inhabitants, now known as the Kanembu, and Sultan Selma, who ruled from 1193-1210 A.D., is recorded as the first black king of Kanem.

By this time effective commercial links were in operation with the coastal states of North Africa, and apparently the Hafsid Prince of Tunis is recorded as having received the gift of a giraffe from the 'King of Kanem and Master of Bornu' during the reign of Sultan Selma's predecessor, Sultan Biri.

The Sefuwa armies (consisting of horse and camel cavalry, as well as foot soldiers armed with spears and bows) were commanded by a *Kaigama* (supreme commander), and the provinces were ruled by governors of whom the *Yerima* (who ruled the lands of western Bornu) was the most important. Unfortunately, some of these provincial governors of Kanem not only engaged in forays intended to enhance the power and dominion of Kanem, but also organised revolts against the sultans, a situation which eventually resulted in such weakening of the administrative structure that the collapse of the Kanem Empire in the fourteenth century was inevitable.

At the same time, increasing desiccation of the land in Kanem was driving many Kanembu southwards in search of better pastures, in the greener lands of Bornu.

## Chapter 7 - Conquest, bloodshed and tyranny

As these people penetrated further and further south, they continuously subdued and absorbed the local tribes, coming gradually to dominate Bornu both linguistically and culturally and to be known as the Kanuri tribe. Ethnically and linguistically, however, the Kanuri comprise innumerable sub-groups and dialects which are distinguishable even today, but which defy all attempts at concise analysis.

Today most Kanuris claim to be able to communicate direct with all other Kanuris in spite of the range of different dialects, as well as claiming to be 'the' people of Bornu. Nevertheless, upon close enquiry, a Kanuri frequently claims to be not only a Kanuri, but also a Kanembu, and possibly also to have ancestry in one or more ancient native tribes of Bornu.

Throughout the remaining chapters I have therefore referred generally to the 'Kanembu-Kanuri' when implying all the people of Sefuwa descent currently living around Lake Chad in both Kanem and Bornu, and separately to the 'Kanembu' and 'Kanuri' when relating them specifically to their own present-day geographical areas of Kanem and Bornu.

With the weakened state of the leadership of the Kanem Empire by the end of the fourteenth century, it is not surprising that an opportunity was seized by the Bulala (either a discontented subordinate clan, or possibly a tribe of neighbouring nomads) to oust the Sultan, then Umar bin Idris, his court and the ruling clan, (the Magumi) from Birnin Jime.

So about 1393 A.D., Sultan Umar and the Sefuwa rulers withdrew to the west of Lake Chad and settled near the Yobe River, but it was actually Sultan Ali Ghajedeni (c. 1470-1500 A.D.), the forty-eighth monarch of the Sefuwa dynasty, who built the first fortified capital and established the powerful Kingdom of Bornu.

This new capital was built of burnt brick near the mouth of the Yobe River and called Kasr Kumo (phoneticised by the Kanuri to Birnin Gazargamo)*. The ruins of this city are still visible today.

Sultan Ali is said to have reconquered Kanem, re-establishing trans-Saharan commerce, and waged war on the Hausa States to the west, subjugating many of them. A series of distinguished sultans followed Sultan Ali, maintaining the sovereignty of Bornu until its defeat (during the famous Songhai invasions from the west and north into Hausaland), at the battle of Nguru in the early sixteenth century.

It took most of the remainder of that century for the Kingdom of Bornu to re-establish its power, but at last (after an irritating delay to his promotion caused by the seven-year occupation of the Bornu throne by a Sultana), one Idris Alooma became Sultan. He unified and pacified Bornu, defeating the southern tribes of his empire - the

---
* Tr. 'Fort Calabash'

Marghi, Gamerghum, and Mandara - and establishing control over the ever-dissident Wadai (of the original Kanem Sefuwa) in eastern Bornu.

The following two centuries seem to have been embellished by few outstanding conquests or other developments, although the ever-present desert raiders, the Tuareg, the Tubu and the Fulbe (Fulani), continued to engage the Kanuri in skirmishes and minor battles from time to time. The overall pattern, however, suggests a period of comparative stability and the self-assured satisfaction of power and influence, during which traditional ceremonies, characterised by their pomp and magnitude, became regularised and established. These ceremonies still persist today and are enacted during the main Muslim festivals (see ch. 9).

This state of complacent stability was, however, to be rudely shattered in the nineteenth century by two disruptive events. At the beginning of the nineteenth century, the Fulani warrior, Usman (Othman) dan Fodio, declared his jihad, or holy war, on the Hausa States. As a result of the repercussions of this jihad the first Sefuwa Dynasty came to an end in Bornu. Then, towards the end of the nineteenth century, the brutal and infamous Sudanese brigand Rabeh cut a swathe of slaughter and pillage into and across Bornu, eventually subjugating the entire Bornu Empire to himself.

Soon after Usman dan Fodio had conquered the northern Hausa states, and had founded his Sokoto Empire, another Fulani force in 1809 attacked Kasr Kumo, driving out the Sultan and his followers. This Fulani force was, however, successfully repulsed by a Muslim teacher from Kanem, one Malam Faki Mohammed el Amin el Kanemi, who commanded his own private army, and came to the aid of the Sultan. The Sultan was thus enabled to re-occupy Kasr Kumo until the dry season of 1811-12, when the Fulani again attacked, this time sacking the city.

The Sultan retreated southwards and set up his new capital at Birnin Kabela, while Malam Mohammed, who now called himself Shehu (Sheik) Laminu, took up residence with his private army in nearby Ngurno on the shore of Lake Chad.

Shehu Laminu was clearly a very brilliant warrior with outstanding military acumen, as well as being a devout Muslim. He faithfully upheld the title of the Sultan, while he himself gradually assumed more and more administrative power. His ability and sagacity were lauded in song and written prose far and wide across the Sudan, and even as far as Europe, and, indeed, some went so far as to attribute supernatural powers to him. Even today his memory is revered, pilgrims visiting his grave to pay their respects.

Shehu Laminu decided he would read through the Koran from the beginning to the end, and that wherever he happened to be when he finished reading, would be the

## Chapter 7 - Conquest, bloodshed and tyranny

site of the new capital of Bornu. It happened that he was sitting in the shade of the *kuka* (boabab) tree which still stands on the west side of present-day town of Kukawa, when he finished reading the final words of the Koran. Here he founded the city of Kukawa in 1814, moving there from Ngurno as soon as it was habitable. It was here that the explorers Denham and Clapperton arrived in 1823, and remarked the double regency of Bornu under the ineffective titular Sultan, and the powerful Shehu Laminu.

Describing their visit, Denham wrote: 'We entered Kouka (Kukawa) and were met by an army of cavalry 4,000 strong under perfect discipline. Chain armour was worn by men and horses, and there were metal plates on the horses' heads. They advanced at full gallop, and, dividing, cried 'Barca! Barca!', (Blessing! Blessing!). In the market place they saw slaves, sheep, oxen, cheese, rice, earthnuts (groundnuts), beans, indigo, brass, copper, amber and coral for sale and exchange. Denham also commented on the presence of beggars. He noted that the Sheik (Shehu) wielded the real power, leaving sovereignty to the Sultan, an eccentric monarch who never showed himself except through the bars of a wicker cage near the gate of his garden 'as if he were some wild beast'.

The fashion in turbans currently favoured at the time attracted comment from the explorers : 'Some exquisites had their stomachs so distended and prominent that they seemed literally to hang over the pommel of the saddle, and in addition, fashion prescribed a turban of such length and weight that its wearer had to carry his head on one side.'

Shehu Laminu died in 1835 and was succeeded by his ambitious son, Umar (Omar), who finally abolished the sultanship in 1846 when Sultan Ari, who had held office for only a few months, was killed in a battle against Shehu Umar. Shehu Umar now declared himself and his heirs to be the only true rulers of Bornu, thus establishing the Second Sefuwa Dynasty in the history of the Kanem-Bornu Empire.

Only five years later, in 1851, during the somewhat indifferent period that followed, Dr Henry Barth came to Kukawa in the course of his explorations, and another thirty-one years later, in 1882, the Kingdom of Bornu tottered under the ruthless attacks of the vicious and brutal Rabeh only to be rescued and propped back into position again, in 1900, by the fortuitous arrival first of the French and then of the British.

Rabeh, brigand of the Sudan and tyrant of Bornu, was born about 1840 at Sennar on the Blue Nile at a period when the Sudan was divided into districts which, in turn, were rented out to merchants who exploited and ruled them. The Bahr el Ghazal Province was the biggest hunting ground of these merchants, and, here, one Zubeir Rahaman was the most powerful. His main camp was at

## Chapter 7 - Conquest, bloodshed and tyranny

Dem Zubeir whence his armed bands of raiders set out in search of slaves and ivory. Thus it was that Rabeh, as a boy, fell into Zubeir's hands. Rabeh grew up to become one of Zubeir's armed followers or 'Bazingers' and in 1863 was selected by Zubeir to command the escort he provided for the Dutch lady explorer Alexandré Tinné during her visit to the Bahr el Ghazal.

Zubeir's fortunes, however, did not follow the pattern of power he desired, and he suffered enforced detention in Cairo, his son Suleiman succeeding him in Dafur and the Bahr el Ghazal. Suleiman, however, did not retain this power in Dafur and was bluffed into confining his activities to the Bahr el Ghazal where he recommenced slave and ivory raids, later resisting government forces sent against him, and finding himself in consequence declared a rebel. All this time, Rabeh served under Suleiman, but when Suleiman was subsequently offered terms of surrender, and prepared to negotiate, Rabeh lost confidence in him, called up his own troops with his elephant-tusk war horns, and deserted to the south. Thus he set off on his own career of sacking, pillaging and slaughtering in sweeps across the southern Sudan even down to the Ubangi. At one time during this period, his followers are said to have numbered 15,000 armed men.

In 1882 he came to the Chari River. Here he made an alliance with one Hayatu, an emigrant from Sokoto, and with him attacked Bargirmi, a vassal state of Bornu.

*Lake Chad versus The Sahara Desert*

# PLATE 21
# TRANSPORT, LIVESTOCK AND MARKETS (A)

1. *(above left).* A mammy wagon: loaded with rolled-up grass *zana* mats. It may take passengers in addition. 2 and 4. *(above right and below right).* Camels may be white, tawny or skewbald.

3. *(below left).* Camel trains still cross the Sahara with merchandise. Food reserves in the hump become very depleted on these journeys.

5 and 6. *(above).* Horses and oxen are widely used as riding animals for individual transport. The red Bornu and white Fulani oxen are both used also for draught work and for carrying loads. *Note;* the *Calotropis* (Sodom apple) plants also in 6

# PLATE 22
# TRANSPORT, LIVESTOCK AND MARKETS (B)

1. *(above left).* In Maiduguri Market in 1955. 2. *(above right).* Beyond Baga: a makeshift market in 1974 near the shore of the receding lake, bartering imported items in exchange for fish and grass mats.

3. *(above left).* Grass mats are a valuable cash/export product.
4. *(above right).* Natron blocks are carried by camels to the local markets from the northern parts of the Lake.

5. *(above left).* The ceremonial 'jester', mounted on a black horse, wearing the head of a ground hornbill (said to be a very wise and discerning bird with special insights) on his own head.
6 and 7. *(above and below right).* Horses dressed in elaborate robes representing traditional body armour: a large padded 'coverall' and silver neck plates to protect against spears and arrows.

## PLATE 23
## TRANSPORT, LIVESTOCK AND MARKETS (C)

1. *(above left).* The large water hole formed by hot water from the artesian bore hole at 'Minetti', our expedition base 1969/70.
2. *(above right).* Animals cluster in a tiny patch of inadequate shade.

3. *(above left).* Hundreds of livestock are driven in from miles around for their daily drink. 4. *(above right).* Sick, elderly and weak animals which cannot reach the water holes do not survive.

5 and 6. *(left).* A farmer draws water with a traditional *shaduf:* a method that conserves water resources. If restricted only for his own uses widespread erosion is also prevented.

# PLATE 24
# DROUGHT 1973/74

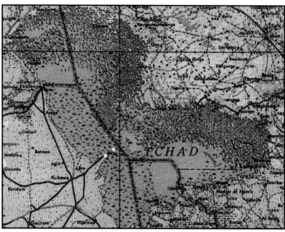

1973 was a year of severe drought right across northwest Africa. The map has been annotated with two white spots to show the distance northeast of Baga to which the lake shore had retreated and where a new temporary port called Dorowon Gowon had been erected. (Plain blue: open water; dark dots: islands).

2. *(above).* Papyrus dried out and died off.

3. *(above).* Fishing became intense in the shallow, shimmering waters.

4. *(above).* The *Merry Mermaid* and inflatable dinghy drawn up on the foreshore of a dune island.

5. *(above).* Heavy boats built of imported timber now filled the foreshore of Dorowon Gowon.

## PLATE 25
## 1974

1. *(above).* Hand cut channels through floating vegetation were essential for access to open waters.

2. *(above).* Hand cut channels through ambach.

3. *(above).* It was very hard work poling through uncut vegetation.

4. *(left).* A square-rigged cloth sail became a real asset for moving heavy canoes in the channels.

5. *(above).* Yedina camp on a very arid dune island.

6. and 7. *(above left).* A fishing barrier set in the Yobe River, and *(above right).* A huge fish trap to be set in the dam. Note the depth of the water there in 1970. The Yobe has not flowed into the lake since then.

# PLATE 26
## SATELLITE VIEW: SHRINKAGE OF LAKE CHAD

1.1963. *(above left).* 'Great Chad': open water at 283m above sea level stretching from northwest of the Yobe River, across the Great Barrier and southeast across the South basin with wide inundation southwards.
2. 1973. *(above right).* 'Medium': open water reduced to 2 basins

 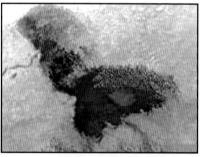

3.1987. *(above left).* North Basin almost dry, little water in S. Basin.
4 1997. *(above right).* Complete desiccation N.Basin; Gt. Barrier dry.

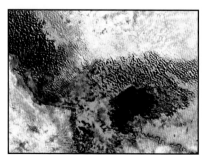 

5. 2001. *(above left).* Desertification increasing all around the lake.
6. 2002. *(above right).* No apparent increase of open water S. Basin.

*(Images above courtesy of NASA Goddard Space Flight Center)*

# PLATE 27
# JANUARY 2002 (A)

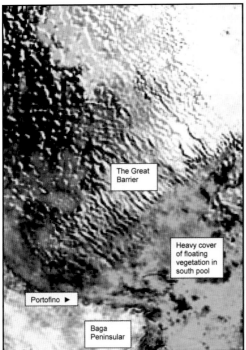

1.1. *(left).* The Great Barrier, showing the NW/SE alignment of the old dunes across it, and irregular dunes in the North Basin. No water is currently being delivered by the Yobe River to replenish the North Basin.

2. *(above).* The Hadjir el Hamis rocks: the only rocks to be found in the vicinity of modern Lake Chad, situated north east of the Chari. River delta, here seen in the arid sahelian conditions now prevailing.

# PLATE 28
## JANUARY 2002 (B)

1. *(above left).* Dune islands in the northern archipelago. 2. *(above right).* Dyke separating cultivated polder (dark green bottom left) from lake water (dark blue).

3. *(above left).* Bol peninsular. 4. *(above right).* Bol town with cultivated polder (green across photo) and arid sahel (top left corner of photo).

5. *(above left).* Water tank on the road from N'Djamena to Bol, with solar panels providing power for the pump. 6. *(above right).* Donkeys on the 'motor road', actually just a sandy side track, a short cut, from N'Djamena to Bol.

Now he moved his headquarters to Logone, and fortified the camp heavily. Then, leaving the women, livestock, and supplies here, he advanced with his troops to Jilbe, and thence began systematically raiding Bornu, now ruled by Shehu Hashim.

In 1893, Shehu Hashim marched against Rabeh, placing his army under the command of one Mohammed Tor. Tor made a costly error in strategy when he led his cavalry through the swampy edge of Lake Chad where they were defeated easily by Rabeh's three thousand riflemen. My Kanuri raconteur ended his description thus: 'Rabeh captured Mohammed Tor and had him sewn into a fresh cow's skin. Then he left him in the sun until he dried'!

Shehu Hashim now set out himself, at the head of 6,000 cavalry with some cannon and foot soldiers. The pennants flew bravely from the lances, the drums and the horns sounded, and the chargers and camels advanced in full armour against Rabeh. Rabeh decided to advance this time, although later on he tended to adopt a strategy of waiting and thus drawing his enemies right up to his fortresses or ambushes. With 500 mounted spearmen and 1,500 foot soldiers armed with rifles and two cannons, he made havoc of Hashim's army, forcing them back to Kukawa. They counter-attacked several times, but flnally he drove them right back into Kukawa itself, while the Shehu fled towards Geidam across the sand dunes.

Rabeh raised a heap of 3,100 human skulls in Kukawa during the month that followed, and looted 950 horses, 1,000 camels, 1,000 cattle, 30,000 sheep and goats, 1,000 ivory tusks, 250 bales of ostrich feathers, 7,000 Maria Theresa dollars, 160 rifles, 100 muskets, 70,000 pieces of indigo cloth, 30,000 pounds of red yarn and quantities of gunpowder and bullets. Then Rabeh returned to Dikwa, where, by mid-1894, he had set up his permanent fortified headquarters, and by 1896 had the whole of Bornu, including the Emirate of Bedde, completely in his power and paying tribute to him.

Shehu Hashim's nephew, Kyari, made an abortive attempt to proclaim himself Shehu and lead his father's army against Rabeh, but the attempt backfired and the people refused to follow him. However, a short time later, Hashim, now living in Geidam, was murdered by two slaves while he was at morning prayer and Kyari now succeeded him as Shehu.

Kyari wrote soon afterwards to Rabeh, challenging him to battle, and their forces met at Dumuruwa near Gashegar on the Yobe River. An exciting and gruesome battle ensued during which the Shehu was thrown from his horse when it was startled by a shot and reared. He was rushed out of the battle unconscious, but rallied and again engaged Rabeh in battle, only to be captured by Rabeh who ordered his slave, Abdullah, to cut Kyari's throat.

During the latter part of his rule Rabeh's resident army at Dikwa consisted of 24 companies, including 1,000 cavalry, 200 throwers of the boomerang-knife from Logone, and 3,200 rifles and muskets, as well as mounted spearmen and bowmen auxiliaries. He also possessed 44 unmounted cannon. He drilled his 360-strong bodyguard daily and reviewed his entire army every Friday. The bulk of his regulars were Sudanese, with a smaller proportion of locally enlisted men. The army lived off the land, Rabeh organising the systematic despoiling of the countryside, his armed bands generally returning with abundant provender and captives.

In the European race to colonise Africa, the shores of Lake Chad were divided by treaty between Great Britain, France and Germany, the boundaries being determined by the simple expedient of ruling lines across the best map then available. France was the first of these nations to establish a footing in the area on the basis of this treaty, and in 1895 the French explorer, Gentil, sailed up the Congo and down the Oubangui and Chari rivers in a little steamer called the *Leon Blot,* entering Lake Chad on 30 October 1897. Fortuitously, Rabeh failed to find him, and, being angered by the arrival of Europeans in what he considered to be his personal domain, captured another Frenchman instead, named de Behagle, imprisoning him at Dikwa.

In 1899, Lieutenant Bretonnet with 451 soldiers was

posted by the French to the Chari. On 17 July 1899, Rabeh, with 2,700 riflemen and 10,000 bow and spearmen, attacked them, and the French force was annihilated with the exception of three men who escaped. Rabeh then hanged de Behagle. Gentil now attacked Rabeh's fort at Kouno, killing over 2,000 of Rabeh's troops, but did not catch or defeat Rabeh. By now the French had determined to eliminate Rabeh, and with well-planned strategy, using three adequately armed forces under Commandant Lamy, Rabeh's army was finally defeated at Lakha, a village four miles from Kusseri. Commandant Lamy died in this battle, and Rabeh tried to flee, but was captured and decapitated by a French soldier who carried Rabeh's head back to the camp on a stick.

Undoubtedly had he survived, and had the European colonial conquest of Africa not intervened, Rabeh would have founded a 'Third Dynasty' in Kanem-Bornu. As it was, the Sefuwa Dynasty was now restored and continues to this day, although its seat is now in Yerwa-Maiduguri and no longer at Kukawa.

Although a complete interloper, Rabeh the brigand-tyrant with his violent, hot-tempered, bloodthirsty brilliance, seems not entirely out of place in the conquests of Kanem-Bornu. In his own compound at Dikwa, the portrait is completed: here he kept a harem 300 strong and had a garden, made with wells and irrigation channels, where he grew limes, prickly pears and pomegranates.

And here the story is told of the woman who came to him complaining that one of his soldiers had stolen her rice. Rabeh promptly had the soldier disembowelled. Finding the rice as declared, he commended the woman for her truthfulness, remarking that, had it been otherwise, she would have suffered the same fate as the soldier.

Now everything began to change, for the French killed Rabeh's son, Fad-el-Allah, in battle and made bargaining gestures towards the rightful Shehu of Bornu, Abubakar Gabai. The British also began to appear on the scene with a force commanded by Colonel T. L. N. Morland. Without having to fight anyone in Bornu, they stopped the bargaining, reinstated the Shehu, and supported him with a resident garrison in Maiduguri. In the same year the Germans occupied the Cameroun with its narrow wedge of Chadian waters, only to be ousted in favour of the French in 1919.

Kanem, now under French colonial rule, and Bornu, now under British colonial rule, proceeded to develop over the next half-century along lines characterised by the policies, direction and impetus of their respective overlords. This era culminated with those searching moments in Africa's history, when many of the former colonies and trust territories achieved so-called 'Independence'. At this critical period, Lake Chad was bounded by four separate potential nations, three of which were French-orientated and one British-orientated.

## Chapter 7 - Conquest, bloodshed and tyranny

With characteristically differing degrees of intention and effectiveness the French and British bowed themselves out, leaving the Kanem-Bornu Empire subjugated to an entirely new kind of suzerainty with its future as nebulous as the mirages of Lake Chad.[39]

Long before the known history of the invaders and conquerors began, however, ancient tribes inhabited Lake Chad and its environs, and, apparently, both the Chad Basin and most of the Sahara as well supported considerable early populations. The immediate indigenous fore-runners of the Sefuwa invaders in Kanem-Bornu are now generally referred to as the 'So' (or 'Sau'), although it is not clear whether this name referred to a particular tribe, to a group of tribes, or to a race of people. There are individuals still living in the southern environs of Lake Chad, who claim So ancestry and who incorporate in their ordinary speech certain words said to be 'So' words.

Relics of So weapons and burial urns, some in an excellent state of preservation, have been found. These are mostly of unusually large size and weight, indicative that their users may have been people of unusually large stature and strength. Tradition holds that the So people were giants. Even today some of the Yedina people whom one encounters living actually on Lake Chad, and many of the Kanuri people of Bornu, are unusually tall people of striking physique.

There is evidence that the So people of the southeastern environs of Lake Chad were long ago driven into the lake by invaders, and adopted a way of life confined to the islands and waters of Lake Chad. Whether the So were the ancestors of the present-day Yedina is also unclear, and, from the findings of linguists and ethnologists, apparently very improbable. The So may well have been the forerunners of some of the resident fishermen of the Lake, other than the Yedina, who use various Kanuri dialects and have entirely adopted a way of life that, in many respects, resembles that of the Yedina.

Apparently no reliable evidence whatsoever exists on which to venture a realistic theory as to the true origin and relationships of the Yedina people. Until very recently these 'pirates of the papyrus' lived a life entirely confined to, and dependent upon, the waters and islands of Lake Chad. Their language is apparently free of major dialects, and incomprehensible to the Kanuri people, who say they are incapable of learning it. The Yedina, on the other hand, readily learn both the Kanuri and Hausa languages for purposes of trade. But both the Yedina and the Kanuri emphatically agree that their languages have no similarity whatsoever.[40]

Who then are these Yedina people and where did they come from originally? Where did they learn to make their papyrus canoes and houses? The historians and scribes of Kanem-Bornu insist, without exception, that the Yedina

## Chapter 7 - Conquest, bloodshed and tyranny

are a people apart - 'original people' - they say, who were 'always' on the lake, and who never intentionally came on to the mainland or became involved in any way in the strife and conquests of the people of the shores.

Are there any clues at all? There are wall paintings in the Valley of Kings in Egypt, depicting reed boats and an engraving on an ancient slab from Nineveh. Moses was also hidden in a reed boat as a baby, and fishermen use a form of reed boat on the west coast of Sardinia even today. Then again, in the marshes at the confluence of the Tigris and Euphrates, in southern Iraq, there are (1970) some 15,600 sq. km. of ancient marshes where bulrushes and *Phragmites* reed-grasses are used to build rafts and canoes very similar in style to those of the Yedina. The people who make these are Arabs whose way of life belongs to the Qurna marshes, but these people have developed their social culture and structural building crafts to a level far beyond the grass canoes and the simple huts used nowadays by the Yedina of Lake Chad. And yet, the design of the arched reed frames of their houses is suggestive more of an advancement upon the basic structure retained by the Yedina than of a radically different or independently originated craft. Is this a possible clue?

I made many enquiries as to the origin of the name 'Bornu'. Without exception, my informants volunteered the information that 'Bornu' is the most ancient name of the area formerly covered by the greater Lake Chad, i.e. Mega-Chad - including all the country of Kanem to the Tibesti Range. The meaning? - 'The River (or Flood) of Nuhu (or Noah)'. Another clue? - if so, is there any connection at all between them?

I do not know how the Yedina got into Lake Chad originally, nor why the vast inland waters and marshes of Mega-Chad came to be called 'The River (or Flood) of Noah' by the ancient people. But it is fun to half-close my eyes and let the Chadian mirages chase one another across my imagination. Then I may try to visualise in imagination those 'mighty men of old, the men of renown', called the 'Nephilim' (was this 'Neolithic Man'?), who lived on the earth just prior to the Flood described in the Bible (The Holy Bible, Genesis ch. 6:4), and of whose descendants a group reappear in Canaan in the time of Moses. Did they drift away in their reed canoes on the rising waters of the Flood and somehow get carried into the swamps of the ancient extended Chad?

Or was the 'Ararat' of Scripture really a Saharan outcrop, where Noah's ark grounded, and where Ham and his descendants were enslaved by his brothers Shem and Japheth, while some of the more independent freedom-loving members of his family maybe escaped in reed canoes into the flooded firki pans and Saharan oases of the Chad Basin (The Holy Bible, Genesis chs. 8 & 9)?

*Chapter 7 - Conquest, bloodshed and tyranny*

***Yes, we may speculate and wonder, and allow the mirages to fool our reason, but the fact remains: no one today knows where the Yedina came from or how they came to live in the papyrus of Lake Chad.***

# Chapter 8

# The Yedina: pirates of the papyrus

During my early visits to Lake Chad I heard people speak of the 'Buduma' but neither recognised nor was introduced to any individual members of the tribe. While planning my 1969-70 expedition, however, I made numerous enquiries about them, but apart from learning that Professor Lukas of Germany had written a booklet (of which no single copy could be located) about their language and that Gospel Recordings Inc. had made a very short 78rpm record of some hymns and scripture readings, I could discover few reliable facts about them.

Turning to the writings of the nineteenth- and early-twentieth-century explorers, I found that Denham[41] had mentioned the 'Biddomahs, a race who live by pillaging the people of the mainland'. Barth,[42] in 1857, wrote of 'the Yedina, or Buduma. . . . as the Kanembu call them, the famous pirates of Tsad'. He also described those curious plank-craft, like a surf-board (made by them from the ambach tree then plentiful in Lake Chad) which they called *fogo,* as well as their papyrus canoes. He mentioned their dependence on fishing for their livelihood and their use of iron spear-heads manufactured by the mainlanders near Kukawa, having the shaft made from the local *kindil* tree *(Acaca spirocarpa)*.

Lance-shaped spearheads were used for war or piracy, while narrow ones with numerous barbs and square cross-section were used for fishing. He remarked their rounded huts made from lake reeds, quite distinct from those of the Kanembu, and their use of local indigo for dyeing the cloth used for their clothes. This indigo was then obtainable in Kukawa market in balls of about 10cm diameter. He saw their beautifully woven, water-carrying baskets with pointed lids, but noted that their pottery was obtained from the mainland.

Barth emphasised that these people were entirely distinct from the Kanembu-Kanuri. He noted that they claim to have three main sections to the tribe, and therefore supposed that they may have had three separate origins in the north, south and eastern shores of the lake. While my own informants, during my 1969-70 expedition, endorsed the fact that the tribe has three sections, they did not agree that this implied three separate origins on the mainland. I must admit that I did not initially appreciate the force of Barth's mention of the word 'Yedina', because all the people I met who were currently engaged on the lake referred to this tribe as the 'Buduma' and even went to pains to assist me to pronounce the word according to their own version of it. So, for the first three months of the expedition while at the lake, I asked about the 'Buduma', had members of the tribe introduced to me as 'Buduma', was told about the 'Buduma' by my Kanuri informants, and even heard missionaries describe the needs of, and pray for, the 'Buduma'.

It was not until I began to hold tape-recorded interviews with members of this tribe, however, that I ran into difficulty over this word. I had arranged for two 'Buduma' tribesmen, one of whom spoke Kanuri, and one Hausa, to be interviewed through an English-speaking Kanuri interpreter. We struggled along as I asked questions about family life, customs, livestock, fishing and boats, but, whatever I asked, the replies seemed stereotyped and inadequate.

Then, one day, a member of the tribe turned up who could speak a little English and could therefore communicate with me direct. I went over all the same ground again without making much progress, and, as previously, we found ourselves back at the name of the tribe, 'Buduma'. I was trying not to appear bored, for these interviews had been unsatisfactory and wearying in the extreme. But it was at this very point that this man, who called himself Ali (a common Muslim name), provided the key that I needed. He paused before answering, and, as usual, said flatly. 'Buduma is name of tribe. It means grassmans.' There was a long pause while I tried to think

## Chapter 8 - The Yedina: pirates of the papyrus

of another useful question, when suddenly, in a rapid, almost emotional tumble of words, he said: 'But Buduma no be our own name for tribe. Is Kanuri name for tribe because they say we be like animal live in grass all time. They say we be no-good mans because we live in grass like animal. But we no be like animal. We be strong mans: we be called true name Yedina.' I got him to repeat the word several times for me while I taped it, and sometimes it approximated to the sound 'Yettnah', sometimes 'Yed'nah' and sometimes 'Yedinah'. Ali himself spelt it 'Yedana'. As the spelling adopted by most previous writers, who have used it at all, is 'Yedina', I have used this spelling throughout this book.

As we went on together that afternoon I found it was as if a key had turned in a lock and a door of confidence had been flung wide open. From that moment onwards, he began to talk freely, explaining their hatred of this word 'Buduma' as used by the Kanuri. He emphasised that the Yedina had never been enslaved by any other tribe: on the contrary they had taken many slaves from all the mainland tribes in the past and all the tribes had feared the Yedina. Moreover they had never even been enslaved by Rabeh. They had in fact sold Kanuri slaves to him, as well as to some of the desert tribes of Kanem. Every time a Kanuri, or anyone else, spoke the word 'Buduma' it was with a sneer, an attempt to belittle the Yedina because of their age-old hatred of them. Perhaps this young man felt more strongly about this than some of the Yedina, but, from now on, I was to discover what a difference it made to our whole contact with them to begin by recognising their tribal status through using their own name properly and respectfully.

From now on I asked for the 'Yedina' when I wanted to find a good boatman or an informant on tribal custom. When we arrived at an island village or camp, I now asked: 'Are you Yedina people?' or 'Is this a Yedina camp?'. The key worked, and they would laugh with appreciation when approached in this way, the door of friendship and hospitality being flung wide open to us.

I began to grasp just how offensive is the use of the name 'Buduma' to them - perhaps something like the feelings of a German when called a 'Hun' in his hearing by an Englishman, or a modern African when he is termed a 'nigger' or 'bush-man'. The origin of the word 'Yedina' however, completely eludes me, as it has done all previous enquirers. Guesses have been made, but they do not really stand up to scrutiny because they are not really based on tangible evidence at all.

The three recognised sections of the tribe are called the *Guria,* the *Madjagodia* (or *Majagujia),* and the *Maibulua.* The former live in the northern part of the lake and are characterised by having two horizontal marks beside each eye. The clans of the Guria include the Mama Guria, the Magana Guria and the Buia Chilim, and centre on Kulua

Island, while the Majagujia are centred on Kan and the Bulariga Kura Islands, and the Maibulua on Yiribu and Ngaloha Islands.

Temple[43] attempted to locate Yedina origins on the mainland opposite each of these centres and said that the Yedina language is basically So, but this theory is not supported by linguists who have compared known So words, from known So descendants on the mainland, with Yedina words. However, the Yedina language has not as yet (1970) been learned to any degree of fluency by any linguist in a position to assess its relationships. It is perhaps wisest, therefore, to wait before attempting to formulate theories as to the real origin of the tribe and language until adequate evidence becomes available*.

However, there may be some basis for suggesting a link with the mainland of the north, south and east respectively *in more recent times* through slavery, for, while the Yedina often speak Kanuri quite fluently, the Kanuri apparently never learn Yedina. This situation may very probably have arisen through the introduction of Kanuri women as slaves into the Yedina villages, the children learning Kanuri from their mothers, as well as Yedina from the tribal group. Both the Kanembu-Kanuri and the Yedina agree that the Yedina have never been enslaved, and that no Yedina descendants can be found in any Kanuri household.

---

* GAUDICHE, Capt. *La Langue Bouduma*. Soc. Africanistes. pp 11-32.

I was quite unable to find out what the Yedina did with the male slaves they captured: perhaps they sold them all to desert nomads. Apparently they did not teach them to fish. It is just possible that they may have used them to herd cattle for them, as one of my Yedina informants rather shyly told me that, during the dry season, when all the able bodied men and women go and live in the floating camps, the old men and 'some people' stay behind and herd the cattle. Asked about these, he said they are not paid. They 'belong' to the village, but could not be described as Yedina - just 'some people'. These may indeed be descendants of slaves. There is also no evidence at all that the Yedina practise any form of slave trading today as do some of the other tribes of the desert regions around.

It is also not entirely clear at what point in the history of Lake Chad, the influence of Islam first began to affect the Yedina whose religion formerly was animist. Sultan Hume, in the eleventh century A.D., had been the first of the Sefuwa kings to embrace the faith of Islam. Then, following Usman dan Fodio's Holy War on the northern Hausa States in the early part of the nineteenth century, Islam swept forward vigorously across the southern Sahara and into Kanem and Bornu, but there is no evidence to suggest that it gained any sort of hold at all on the Yedina until after the 1920s. Every Yedina informant interviewed who was under the age of about thirty-five years, refused point-blank to admit that the Yedina had any religion but

Islam, and was embarrassed by the suggestion that any members of the tribe could possibly be other than Muslims today. Only close probing of older people yielded any useful information about the former religion and that was at best scanty. Yet it is a fact that the majority do not understand the teaching or demands of Islam, and merely pay lip service to it.

It was thus extremely difficult for a casual, short-term visitor to the lake, such as myself, to identify or define the genuine traditional culture and way of life of the true Yedina, and distinguish it from more recent adaptations and adulterations resulting from the *Pax Britannica,* French anti-piracy controls, modern transport communications, commerce, radio and religious proselytisation. Only by complete familiarity with, and fluency in, the Yedina language over a prolonged period could one begin with certainty to elucidate the genuine Yedina religion and culture.

What did emerge clearly, however, from my enquiries, was that the Yedina are indeed a very ancient people, originally completely isolated from the surrounding mainland tribes, essentially 'reed-orientated', if I may coin the expression, yet strongly pastoral, and above all proud, independent, and generally of outstanding physique. Moreover, their characteristic staple diet is a combination unusual in Africa, if not unique: fresh cow's milk and fish.

At least one section of the Yedina tribe, if not all, worships the Spirit of the Lake, performing an annual rite during which a woman prepares 'grain', (of unknown source or purpose) which, as part of a ritual ceremony, is thrown into the lake. If fish come and eat it, the omen is propitious and an abundant take of fish, cattle and grain (perhaps as cattle feed or for soup/gruel cooking) may be expected. Similarly the spirit of the *Acacia albida* tree *(karaka)*, the largest tree of the lake shores, is worshipped, and offerings of fish and cereal soups are poured into hollows in the ground around its roots as propitiatory sacrifices.

They also believe in a terrifying supernatural monster or djinn that lives in the lake, whose skin is pale like that of a Fulani tribesman and which has long thick hair on the head. It is said to reach out long thin arms with which it captures transgressors, dragging them to its lair beneath the water. Although they believe that all supernatural beings originated from the water, they say the first man sprang from the ground, like, or near, a *karaka* tree.

They think that foreknowledge of future events is sometimes communicated in dreams and, symbolically, through certain events and coincidences. They deny the possibility of any communication between the dead and the living, or of any kind of life after death. Possibly they see their union with the Spirit of the Water as a sort of eternal reality. They have specialists and consultants who can, for a fee, handle and advise on the occult.

At death, both men and women are buried in the sand outside the village. Some said they are buried in a sitting posture (like the ancient So) but my younger informants said the bodies are merely washed, wrapped in a cloth, and, with fasting and mourning, simply buried in a shallow grave. The grave has to be shallow in most cases owing to the high water table. Apparently, funerals are attended by every villager in the vicinity at the time. Infant mortality on the lake is extremely high, so an infant is not named, nor is its sex disclosed, for several days. It is thus not even regarded as a human being for the first few days of its life, an interesting, psychologically protective, custom to insulate the mother against the intensity of repeated disappointment and disgrace.

This custom is shared by a number of African tribes. Many Yedina babies and children die of cerebral malaria; and cerebro-spinal meningitis is common. Others die of quick, vicious fevers which may also carry off adults. If the infant succeeds in surviving the first week of life, a naming ceremony is held. At this ceremony the true name is given to the child as well as an oblique name by which it can subsequently be addressed or called. The true name is not used again. The oblique name is usually descriptive, such as 'the son of the brother-in-law of the shield maker', but may nowadays be replaced by an adopted Muslim or Christian name.

Little boys and girls may swim and play in the lake together and around the villages. Small infants may accompany their mothers by boat to the floating camps, and children of only three and four years of age learn the movements of poling and paddling a canoe. Even small children are to be seen with their fathers tending the nets and sorting the catch, and by the age of seven or eight they can handle quite large canoes by themselves, developing wonderfully strong little bodies.

It is the work of the women and girls alone to weave the milk churns and special water-holding baskets with pointed lids, reclining mats, mosquito-proof mats, and the houses, both of the camps and dune island villages. Both men and women assist with cultivation (cultivation other than of pumpkins being almost certainly a relatively recent addition to Yedina culture), but only men herd and milk the main herds of cattle, while women must care for the calves. Although the women can and do paddle and pole canoes, the men build them, and do the actual fishing. The making and care of the fishing nets and lines is men's work, while the preparing, drying and cooking of the fish is women's work. Near the commercial markets, however, where the men have stepped up their fishing output above their own subsistence level, I noticed that much of the preparation and drying is also done by men, although in most cases, on enquiry these turned out to be lake Kanuri or immigrant Hausa and not Yedina *(Plate 16:1 and 2)*.

Boys are usually circumcised between the ages of fifteen and eighteen years, in groups. They spend the circumcision period in a special compound of reeds just outside the dune island village where they are attended by the local practitioner for a month. Yedina boys nowadays (1969/70) do not marry until they are between twenty-five and thirty years old, although they are permitted free intercourse with girls before this. However, I feel dubious about the authenticity of this custom, and suspect that it has been introduced from the Kanembu-Kanuri in recent times, as the older Yedina say it is a very terrible disgrace for the boy to sire a child prior to official marriage, and in any event he is absolutely forbidden any premarital sexual relationship with the girl destined to be his first official wife.

Since this wife is chosen for him, and the interparental bargaining completed without him; since she is presented to him as his bride immediately after her second menstrual period following forty days of supervised isolation from other people, in company with one 'grandmother'; and since her genuine virginity is prized greatly by both the suitor and his parents, it seems most probable to me that the modern laxity in the arrangements is due to the recent super-imposition of Islamic and Kanembu-Kanuri customs.

Under Islamic custom, divorce is easy and frequent and there are always large numbers of unmarried (divorced) girls and women about, with whom the unmarried young man may legitimately have intercourse. He may not, however, in any circumstances, commit adultery. If he becomes the father of a child out of wedlock, it is a simple matter to absorb the child into the parental, or his own future, household. A virgin is usually chosen for him if possible for his first official wife as she is young and submissive and can be trained, from the start, to his ways. Subsequent wives can be divorced women, and in any case, if he is important enough, he will wish for the additional assistance of concubines whose services he will reward by fathering their children and retaining them as members of his own household - or, in reality, as his servants.

I noticed some confusion in the descriptions given by my younger Yedina informants, for it was apparent that the Yedina do not really favour extensive polygamy at all, and even as many as two wives are not universal in the tribe. Moreover I sensed a sort of embarrassment when one of my informants explained that some Yedina men go so far as to find it 'convenient' to have one wife in the floating camps, and one left at home in the dune island village.

It seemed to me that they would have preferred it if Muhammad had not suggested as many as four as the prescribed upper limit on the number of legitimate wives.

*Chapter 8 - The Yedina: pirates of the papyrus*

And as for concubines, I got the impression that they were not really in style at all among the true Yedina, and that the idea was really an Islamic intrusion into this tribe. As my enquiries progressed, I found myself wondering if the Yedina were originally monogamous. It is rare to find African tribes that do not regard the multiplication of wives as something of great status value. No one I have met seems to know the answer. However, it was quite clear that my younger informants had, in practice, embraced the Kanembu-Kanuri-Islamic customs in these matters, and appeared to show no embarrassment whatsoever over describing their more intimate relationships and exploits to me.

When the young Yedina man marries his first official wife nowadays, he presents a length of cloth to *her* father, and the girl presents fresh milk in a woven vessel to *his* father. The bride price for a Yedina virgin at the present time is said to be round about N£30 (the same as the price of a large dugout, wooden canoe!) plus at least four cows. Before the wedding ceremony, the groom's female relatives weave a new hut with a sort of annexe or verandah leading into it. The people gather at the village for drumming, dancing and feasting, a popular ancillary entertainment being wrestling. These curious people are said never to have included the brewing or drinking of any kind of intoxicating drink in any of their ceremonies or tribal life, fresh cow's milk alone being the accepted drink.

Nowadays, however, Yedina who trade in the markets buy and drink bottled soft drinks and beer, as well as chewing tobacco. At some defined point in the marriage ceremony the couple go to their house and an old 'grandmother' settles down in the annexe as observer. As soon as she is satisfied with the evidence vindicating the girl's virginity, she hastens to the groom's parents, and indeed the whole village, proclaiming the good news. The festivities are said generally to continue for as long as a week.

In cases of tribal disaster, like drought or tornado, the people of neighbouring islands and villages are called together by drumming, and may carry out propitiatory ceremonies. Here again, in answer to my questions, I encountered considerable confusion in my informants' minds as to whether these were animistic or Islamic in character, possibly because they had simply never learned where the one religion ended and the other began.

The commonest cause of all adult deaths in this tribe is said to be drowning, especially during the dry season months when most of them are out on the lake, and when the strong gusty squalls of February, March and April strike their camps and boats. It is also not uncommon for people to be lost when a floating *kirta* fragments, and part floats away with someone on it.

## Chapter 8 - The Yedina: pirates of the papyrus

Almost invariably when we approached a dune island in the *Jolly Hippo,* a group of villagers would come down to the shore to meet us. If we anchored off-shore they would come out in their canoes and either escort us, or bring us with them, in their own canoes, to their island. Then we would be led to the house of the village headman, and mats of sheep, goat, sitatunga hide or woven grass would be spread in the shade for us to sit on. Sometimes a Hausa-speaking Yedina could be found to interpret for us, but more often communication had to be by means of signs and mixed words of French, English and Hausa. In this way we managed to communicate with a surprising amount of success, so long as we stuck to subjects of immediate interest such as our whereabouts, the well-being of the cows, fishing and suchlike.*(Plate 5.2).*

They were generous with welcoming gifts of fresh milk, milked straight from the cow while we watched, and fish. As natural boatmen, they always liked to come aboard the *Jolly Hippo* and look inside, thoughtfully discussing its merits and demerits with one another as compared with the government and Mission vessels and their own canoes.

The *kadai (Plate 13),* or papyrus canoe, is the traditional Yedina craft. This is constructed of papyrus stems bound together with rope which is woven from the fibres of the leaves of the doum palm, and then soaked in water. The stem of the papyrus plant is triangular in cross-section and its pith is extensively aerated, giving it a natural buoyancy in water. When a new canoe is needed, one or two men go into a stand, or floating raft, of tall papyrus and cut sufficient to complete the size of craft they want. They use a sharp, locally made cutlass about 40cms long, a dagger, or an imported steel machete, and cut the stems neatly through, so that each end is sharp and wedge-shaped. When they have cut sufficient papyrus, they tie the stems in large bundles and then tow them behind a canoe to a shallow, sheltered place with a firm bottom near the village or camp. Here also they lay the rope bundles in the water to soak. Beginning with the prow, they tie the ends of the stems in a bundle of about 10cms diameter so that the other ends splay out freely. The tying of the front end is done very carefully with special kinds of knots. The stems are laid in the bundle so that their flat sides lie against each other *(Plate 14.1 - 6)*.

They then insert, one by one, the wedge-shaped ends of additional stems, from the back, into the open end of the prow bundle. All the stems used may be about 3m to 4m long. These new stems are pushed well home into the bundle and then another ring of grass rope is woven in and around the bundle a short distance behind the first ring. Another set of stems is inserted and the process repeated. Now a short, sturdy, forked stick is pushed or hammered vertically into the ground and the prow rested upon it, in the fork. The prow is stroked into a nice curve and a new ring of binding woven round the bundle, the knots being pulled tight with the builder's teeth.

At this stage no more stems are inserted into the bundle unless it is to be a very long canoe. This one will be about 4m long. For a very long one, longer papyrus stems will be sought and the inserting and binding will be continued longer. The existing bundle is now bound with a series of rings of woven grass so as to form a neat parallel-sided centre bundle which will form the floor of the canoe. When this is complete, they return to the prow and begin to add more stems by pushing them into the sides of the prow bundle, and then additional ones into the backs of the new lateral bundles, until the lower part of the gunwhales has been completed. Another bundle, with its additions to elongate it to the full length of the developing canoe, is then added to each side above the first two.

Some rather specialised work is now done on the prow, perfecting its graceful upward curve by means of additional bands of woven rope, and finally it is stabilised in its permanent position by means of a stay from its tip, tied securely back to a point in the for'ard end of the floor bundle Now the builder goes to the stern and does some extra strong binding and weaving right round the whole body of the canoe, finally trimming the long ends of jutting papyrus with one or two clean strokes of his machete, so as to leave the stern square and clean, with the sectioned ends of the cut papyrus gleaming white.

Immediately the canoe is completed, it must be launched or the papyrus and palm rope will dry out and become loose. It will remain in the water at all times and may continue in use for at least two years. *(See Plate 14: 7)*. When launched, a new canoe rides high in the water, but as time goes on it gradually subsides due to the entry of lake water into the fronds by osmosis. However, these canoes are completely unsinkable, and if they seem to be rather damp inside, all that is necessary is for the owner to add a bundle or two of papyrus stems along the floor, and perhaps build up the gunwhales a bit.

When a really big canoe is needed, as for transporting cattle, it is constructed in exactly the same manner, the only difference being the use of longer papyrus stems and more bundles. When the Yedina need to live and sleep in their canoes for some days, they sometimes put a flat grindstone (imported from elsewhere through the local markets, as no stones occur near Lake Chad, except at Hadjir el Hamis) in the bow, on the floor. This serves as a base for making a cooking fire. On the floating islands however, fires are lit on platforms of green reeds, built up on the rhizome mat of the island.

## Chapter 8 - The Yedina: pirates of the papyrus

In the past *kadais* were paddled with a round-ended paddle. Nowadays they may be either paddled as previously or poled, using a long palm frond imported from the Niger and Benue rivers, and sold in the local markets at about sh. N5/- per frond. On the Benue River, they are on sale at twenty poles for sh. N1/-. Benue-style, spear-shaped wooden paddles are also imported now to Lake Chad and sold in the markets, but one does from time to time still see the old-fashioned round paddles.

Ambach was, in the past, as much a part of Yedina life as papyrus. The famous ambach plank-rafts are now scarce and rarely seen. Formerly each Yedina household possessed at least one or two of these *fogo,* men, women and children being equally adept at their use. They would lie full length along the board, which may have been any length up to 7m, and which was feather-light with a density of half that of cork, and then paddled or 'swam' rapidly, using hands and feet.

In the early days of the tough line taken by the French to control lake piracy, the Yedina used to elude the French patrol boats by leaping on to their *fogo,* and plunging into the water between islands, paddling hastily across to the other side.

Then, with the *fogo* over their shoulders they ran across the island to the other side, plunged in again, and quite literally ran rings around the French patrol boats, until shots would be fired and the 'culprits' (or their remains) dealt with appropriately.

The Kanuri believe the Yedina to be amphibious. It is true that they can remain under water for astonishingly long periods, and never hesitate to jump out of their canoes and push them into the reeds if a gale or other danger blows up. We were amused on one occasion when we were at the north end of the lake, on a westerly course among the islands east of Nguigmi, within the waters of the Republic of Niger: as we rounded an island, we saw a canoe rounding another island ahead of us and coming towards us.

The moment the canoeists sighted the launch, they turned about and paddled frantically towards the nearest reeds. Watching them through the binoculars we saw them leap into the water and swim under water, pushing the canoe hastily out of sight, and then, with only their heads showing at the surface of the water, watch us furtively through a thin screen of reeds. We later learned that these may have been tax or customs evaders who supposed the *Jolly Hippo* to have been a government vessel.

Ambach was also used in the past for the manufacture of shields. Narrow ambach planks were laid across a hollow in damp sand and then covered with wet sand,

## Chapter 8 - The Yedina: pirates of the papyrus

so that they assumed a nice curve matching that of the hollow. The strips were then stitched together with palm fibre and decorated by a special shield-decorating artist. So fine was some of this carving that early writers described the shields as resembling engraved bronze.

As mentioned earlier, spearheads for fishing and for piracy, daggers, and grindstones were all imported from the mainland. I do not know who the middlemen were in these transactions for it seems most unlikely that the Yedina would have walked into Kauwa market or Kukawa to do their weekly shopping in the old days. It seems much more likely that these goods were bartered by the Kokoto tribe to the north, and desert nomads, in exchange for slaves. It is just possible that some of the lake Kanuri, who had begun to adopt a way of life dependent on fishing, acted in more recent years as the middlemen.

On many of the dune islands and along the north-eastern shores of the lake, the houses are sheltered or united by reed fences stretched between them, forming a sort of compound. I got the impression that this also is probably a reflection of Islamic influence - with its need for 'closed' compounds and households - rather than ancient Yedina tradition.

The Yedina houses are built entirely of reeds laid over arches made out of reed bundles, bound with the usual doum palm fibre rope. The whole consists of a single, beautifully woven unit, and can, in the event of a severe rise in lake level, be lifted bodily by only a few men and carried to higher ground. The Kanembu-Kanuri type beehive hut, in contrast, is a much bigger, heavier affair, made of shore grasses rather than light-weight lake reeds. It takes about fifty men to carry a Kanembu-Kanuri type hut, and is a major undertaking *(Plate 17.3 - 5)*.

Life on the dune islands is centred in the cattle rather than in cultivation, although millet, maize and wheat are now popular on some islands and on the north-eastern shores, but crops are still not grown at all by many of the Yedina. The traditional cattle of the Yedina are large white or grey cattle with enormously enlarged bases to the horns. The actual bone in the horns contains large, air-filled cavities, and the horn covers this in the normal manner, but with a distinctive shape. Nowadays, this line of Chad cattle has been adulterated by the introduction through trade with the Fulani of the *white fulani* and the *red bornu* strains *(Plate 16.3-4; 21.6)*.

Although I have heard livestock experts at the Institute for Agricultural Research in Zaria, Nigeria, hotly deny that the horns are in any way related to the swimming habits of the Lake Chad cattle, the Yedina state categorically that their performance in swimming long distances is definitely better than those of other breeds, and attribute it to the horns.

## Chapter 8 - The Yedina: pirates of the papyrus

Personally I can see no point in trying to deny something which is completely obvious to the people who handle the cattle daily. The explanation, as is so often the case, probably does not lie in calculations of bone density or absolute head weight, but rather in the fatigue element. There is absolutely no doubt at all that the head of a Lake Chad *Kourri* cow, swimming, lies more comfortably on the water surface than does that of the cow of other breeds, so that, for long distances, or in rough water, it needs to expend less energy than if it lacked this extra buoyancy device.

We found evidence of the former presence of cattle on dune islands far out in the lake, but whether these had swum there or been transported by canoe we could not discover. Apparently the Yedina do not slaughter their cattle for food, using them primarily as currency, for milk production and, occasionally, for sacrifice. Nowadays many Yedina trade their cattle in the markets for cash.

Unlike the Shuwa and the Kanembu-Kanuri, the Yedina never use horses or camels, but we frequently noticed sheep, goats, fowls and ducks, as well as some wild-caught pets, around the dune island houses.

During the dry season all the able-bodied men, women and children go to their fishing camps on the floating islands, leaving just the elderly and sick and 'some people' to mind the villages. Choosing a suitably sheltered site on a papyrus or *Phragmites* island, the men cut down fronds which they lay horizontally to form a stable platform. Then the women begin to weave temporary huts, building up and packing the floor inside with reeds until it resembles a wide shallow nest. This is the bed.

In the past, woven mosquito-proof mats were hung over the door, but nowadays (1969/70) cotton baft, sewn into a rectangular, tent-like arrangement, can be purchased in the markets and has almost completely replaced the mats. This so-called net is suspended in the new hut, by its four upper corners, the sides hanging down like a continuous curtain. When a men's temporary camp is set up, four reed bundles, with horizontal cross bundles, serving as a frame to hold up the net, are considered adequate, no hut being built at all. A reed mat may be laid across the top of this frame to provide shade during the heat of the day *(Plate 15)*.

The men do the fishing and, in the past, constructed and maintained the nets as well. These were made from fibres of the *Calotropis* plant, and had a rather wide mesh suitable for large fish. They were upheld by papyrus stems or ambach sticks (cut short) serving as floats, and weighted with the bones of cattle. They never used lines as they had no hooks, and never made basket traps. Barbed wooden spears were used then, as now, for spearing large fish caught in the meshes of the net before pulling them out.

Nowadays these old locally-made nets have been entirely replaced by expenslve imported nylon nets (of various gauges) and nylon lines laden with imported metal fish hooks manufactured by a well-known British firm. Although large fish may still be caught, vast numbers of smaller fish are also netted, dried and smoked, and then exported daily through the mainland markets *(Plate 12.1).* In the past only sufficient fish for home use was caught, this being eaten raw, sun and wind dried, smoked, roasted or made into fish soup. The need constantly to ship the dried, smoked fish to the mainland markets has also, naturally, brought with it something of a change in the habits of the fishermen, and the development of a thriving freight-canoe service across the lake and among the islands.*(See Plate 12: 3).*

The fishermen usually (the details of their practice varying with the season and the moon) set out from the camp for their fishing grounds in the afternoon or late evening to lay their nets and lines, continuing this work on into the night if it is full moon or thereabouts. They take up the nets again in the early morning so that the fresh catch is landed ashore, or at the camp, before midday. Then the fish are immediately scaled, washed, prepared and put to dry by the women. In some cases they are hung on lines for sun/wind drying, and in others they are coiled round, or else cut into chunks, and laid on reed racks in the sun above smoky grass fires. These fish-drying camps are often beset with thousands of flies, and attended by kites and vultures. *(see Plates 11.4 and 5; 12.1)*.

In former times, fish was selected for palatability, but nowadays every species is lumped together indiscriminately on the racks, and sold, ungraded, by the sackful to the traders who ship it to city markets, far from the lake, where it will not attract the scrutiny of connoisseurs. Indeed, the dried fish from Lake Chad usually arrives at its southern markets after three or four days' journey by lorry through the heat of tropical Africa, enriched by the thousands of fly maggots which have hatched out during the journey, a fact which apparently does not trouble the average city consumer!

How piratical are the pirates of the papyrus today? To us they were charming, hospitable, friendly and generous, just as long as we were identified as having no conceivable connection with the tax and customs collectors or the Lake Police. The only two occasions when I encountered them in a different mood were when I was searching for my missing yacht, and was then aboard the Mission launch, and on the occasion when the canoeists at the northern end of the lake fled into the reeds at our approach. But in the latter case, we were not even sure that they were Yedina people.

I admire and like the Yedina. I found them to be naturally self-assured, extrovert people, well-integrated, even nowadays, (1970) as a tribe. They are physically tall, powerful and intelligent - masters in their own environment. They are intolerant of rivals, invaders and intruders, especially when the latter, under the cloak of indigenous neo-colonialism, arrogate to themselves the role of overlords, referring to them, with that special curl of the lip (as if speaking of ignorant and inferior beings) as 'grass-men', 'Buduma'.

I had the feeling that the Yedina are sometimes sad today that their age-old methods for dealing with such people can no longer be practised. They would certainly be at a disadvantage without their ambach shields and fogo, and their war spears.

*Lake Chad, its papyrus, and its ancient Yedina people belong together; a fragile eco-system in a very insecure modern world.*

# Chapter 9

# Fish, livestock and markets

The introduction of nylon gillnets and lines, together with factory-made fish-hooks, in the early 1960s, has revolutionised fishing as an industry on Lake Chad. These have already completely replaced all native-made fishing gear on the lake, although old-fashioned, locally-made, reed and wicker dams and fish traps are still used on the four feeder rivers. Cast-net fishing is still practised on the Chari, but the home-woven nets have been replaced by nylon. Since gillnet and hooked-line methods of fishing are independent of the type of boat used, kadais are still appropriate and common on the lake, although larger imported canoes and plank-built boats have been introduced in recent years. Even the simple Hausa 'gourd-boat' can be used for both gillnet and hooked-line fishing.

The gourd-boat is really just an enormous gourd (calabash) with an opening of between 15cms and 20cms diameter at the top. Instead of using a conventional boat or raft, the fisherman lies on top of the gourd with his stomach over the opening, and his fishing spear over his shoulder (held in the fold between his neck and shoulder). Then, using hands and feet, he paddles along looking like some enormous kind of spider. In this way he lays his lines and nets (carried to the site inside the gourd), and then returns later to inspect them. If he finds a small fish in the net or on a hook, he simply disentangles it, and, neatly lifting his stomach to one side, pops the fish into the gourd. If he finds a large fish, he spears it, before removing it from the net or line, and pushes it into the gourd. If he gets a real 'whopper' he is in a bit of a spot, but can generally, after killing it, carry it home on his head while paddling his gourd. Naturally the gourd-fishermen are confined to the calmer waters of the lagoons and the Yobe and El Beïd rivers, when they are low *(Plates 25.1;12.4-5).*

The gourd-rafts that Edith Jackson, Mohammedu and I used in 1955 are similar in principle to the gourd-boats for buoyancy and as receptacles for fish. They are also more suitable for less turbulent waters. Some unusual and somewhat specialised traditional wicker traps used in conjunction with fish dams constructed of sticks and reeds, as well as gourd-boats may still (1969/70) be seen on the Yobe River *(Plate 25.6-7).* These are used at certain seasons of the year for catching fry, but wicker traps are not generally used on the lake itself. A gillnet is a curtain of netting, suspended vertically in the water, upheld at the top by floats of ambach or cut papyrus, and weighted at the bottom by bone sinkers.

When fish swim about in the water naturally, they may swim into the net, their heads going between the meshes. If the mesh size is appropriate to the fish, the fish will be able neither to get its whole body through nor to withdraw its head because its gill covers now get entangled. A fish hitting the net with great force may also become completely wedged in the mesh.

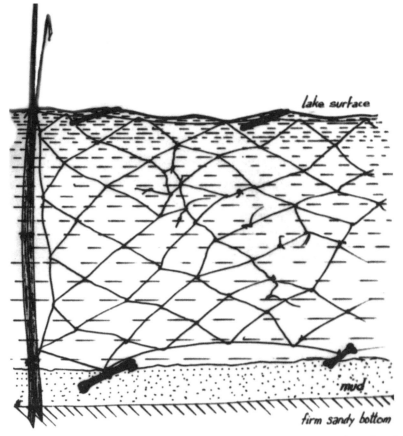

Diagram 21. A nylon gillnet at Lake Chad in about 1½m depth of water. Each end is attached to two papyrus fronds, ambach or papyrus floats lie along the surface and animal bones serve as weights at the bottom

A gillnet also acts as a sort of sieve or selector, permitting small fish to pass through, while fish too large to get their noses into the mesh cannot get caught. Repairs are effected by simply knotting broken threads.

Hooked lines are just long lines upheld by floats, and loaded with barbed hooks dangling at intervals on short lengths of thread. These may or may not be baited. Random fishing by means of lines carrying un-baited hooks is termed foul-hook fishing, and in Lake Chad is sometimes the means whereby really enormous fish are caught. It is said that Nile Perch of over 100kg have been taken in Lake Chad by this means, but they do not seem to be at all common today, and the one shown in *Plate 10.4*, weighing 52kg, was the largest that I saw personally.

Productivity research has covered many aspects of the lake fisheries, including fishing technique and seasonal variations in distribution, quality and quantity, of catch. It is a little confusing that mesh and twine definition in the English and French-speaking countries differs. In Nigeria, nets are sold in the markets according to stretched mesh size (given in inches), and length and depth of net in feet. The French-speaking countries define mesh as 'bar length' measured in millimetres, length being obtained by stretching the net horizontally to its maximum, and depth by counting the number of meshes. Twine size is also important, for not only must its strength be adequate for the largest fish likely to be caught in a given mesh, but also its price rises with its thickness and strength. A 7½ inch stretched (95.2mm bar) mesh net made of twine of 210/18 denier lasts 100 fishing nights in Lake Chad while a 2½ inch stretched (31.7mm bar) mesh net made of twine of 210/3 denier lasts only thirty fishing nights.

Since the introduction of nylon fishing gear to Lake Chad it is extremely simple for a fisherman to develop his business from small beginnings. With a gourd-boat, or, if he is a Yedina tribesman, his home-made *kadai,* and N£4 worth of 4-inch stretched (50mm bar) mesh gillnet, purchased in Baga, Nguigmi, Bol, Wulgo or other markets, he can go out into nearby waters and catch a few good-sized fish each night, drying them and selling them a few days later. Soon he can afford to buy an additional length of the same mesh and join it on to the first piece, or he may try a different mesh, perhaps 2-inch stretched (25.5mm bar), which is especially suitable for *Alestes baremoze, A.dentex* (silversides), and other smaller fish. Soon he buys a large mesh, say, 8-inch stretched (101mm bar), and tries for the larger types, *Labeo spp* (African carp), *Lates niloticus* (Nile perch), *Heterotis niloticus,* and *Citharinus spp* (moonfishes).

With his wife or wives and children helping, his ever-increasing daily catch is prepared, dried, smoked and sold to the traders on the shore. Soon he feels the need for a larger vessel, and buys a Benue River dugout canoe,

## Chapter 9 - Fish, livestock and markets

adding a plank to the gunwhales on each side to enable him to go across the open waters in safety, stay out longer, and handle a larger catch. After a short time he finds himself selling between N£2 and N£5* worth of fish daily, and before long he also owns a bicycle, radio, an extra wife or two, and a stall with a concrete floor, a spacious rear compound and a retail store in front, in the market port. As the business thrives, he goes into partnership with his brothers, opens a petrol station, sets up a flour-making machine complete with generator, and is well on the road to prosperity, in 'Lake Chad 1970s' style.

There is ready money to be gained through fishing on Lake Chad even for a man who comes as an immigrant and starts in a small way. But it is hard work, and, for the outsider from Sokoto or Kano, even if he is already an experienced fisherman in his home area, the lake holds many terrors. Unlike the Yedina he is not born to the lake and its whims. Nor does he navigate by means of maps, compass, and log. To the Yedina, however, and to a lesser degree, the Lake Kanembu-Kanuri, prosperity is on the doorstep, and neither tribe is backward in taking full advantage of this situation. I think it is doubtful, nonetheless, if the Yedina would so willingly have shared the spoils, or tolerated this commercial rivalry, had the French not so effectively curtailed their former piratical tendencies and practices.

---

*Nigerian pounds, which preceded the Naira

It is not altogether surprising that research workers have reported a marked fall in mean catch per unit effort since 1962. In the case of gillnets of 7½ inch stretched (95.2mm bar) mesh there was a drop in mean catch per unit effort over five years, to 20 per cent of its former value. A slightly smaller fall in mean catch per unit effort over the same period was found for 4-inch stretched (50.8mm bar) mesh. This is attributed directly to the increased fishing intensity on the lake.

Most Yedina fishermen use their age-old floats made of papyrus stems or ambach sticks, cut short, to uphold the nets, and bones as sinkers. Fisheries research personnel have tried to introduce the use of cork floats and metal sinkers, but there is as yet no ready market for these and local fishermen claim that they hold the nets so taut that a diminished catch results. Similarly their attempts to encourage the salting of fish in preference to sun/wind and smoke drying have not been generally acceptable to the lake fishermen. A suggestion that nets should be laid in the line of the prevailing wind is also generally ignored, at least by the Yedina, who obviously feel that there are other factors of greater significance affecting the positioning of their nets relative to potential catch.

Similarly, attempts to introduce 'better boats' in the form of plank-built dinghies capable of mounting an outboard motor, designed by advisers from the Food and Agriculture Organisation (F.A.O.) of the United Nations, have met with only limited success. There is little likelihood of these replacing the kadai or dugout canoe, although the smaller plank-type canoe introduced long ago by the French is widely used.

The trouble with the F.A.O. dinghies is, of course, that they are ill-suited for paddling or poling when the motor fails, few fishermen can aflord them (prices range from N£85 to N£200) or the outboard motors, and few are competent or can be bothered to maintain them for long. Moreover these boats ride higher in the water than native canoes and the French plank-canoes, and therefore soon warp and become useless. Wooden dugouts and French plank-canoes are smaller, and, if out of use for any length of time, can easily be submerged so preventing warping and excessive leakage.

At first, when I started to use the local canoes extensively, I was worried over the continual water baling that seemed necessary, but, as time went on, I not only learned that canoes are often deliberately allowed to leak and deliberately sunk, but that the water which so faithfully seeps in through minor cracks is good for the wood. When canoes are brought ashore for repair or for improvement, such as the raising of the gunwhales, they are filled with wet sand and often covered with a mat shelter. I find myself constantly wondering why the fisheries and F.A.O. men did not introduce fibreglass canoes which could accommodate their outboard motors instead of heavy plank-built craft. In the event of the motor failing, they would be light enough to pole or paddle, and, of course, would not warp. They could even be provided with built-in buoyancy units. The cost would at first be relatively high, but I venture to think that they would be genuinely practical and would soon become sufficiently popular to appeal to the fishermen and thus have an assured market.

The District Head of Kukawa was, at the time of our expedition, involved in a very interesting modern trade development in the Lake Chad fishing business: he had organised the people of Baga and Portofino to supply fresh fish (in addition to the usual dried flsh exports) for immediate shipment, on landing, in refrigerated lorries to Kano, 770kms by road from the lake shore. Although I could not obtain data on which to assess the value of this business, it was quite clear that it was proving to be highly profitable.

For the time being, however, the local fisherman thinks of his fishing and processing methods only in terms of the cash paid into his own hand by the trader at the nearest market port. The fact that his product has become filled during its crude processing with the eggs of thousands of flies, which hatch out in the smoked fish during its long journey through the tropical heat by lorry to distant city markets, simply does not interest him.

He may notice that he gets fewer and fewer of the large fish in his nets each year and may regard this as a nuisance, but plenty of smaller mesh nets are available in the markets, and, by using more of these, he can still obtain the same mean daily weight of fish. That he is not only reducing the actual population of the larger fish and interfering with their breeding potential does not worry him because he feels the lake is large and wide and, wherever the fish come from, there will surely always be more. He views the collapse of tribal barriers, at least at the commercial level, as a means of entry to wider markets, and as a situation to be exploited to his own advantage.

Meanwhile, any thought of restricting the fisheries to local, or licensed fishermen, is not yet even mooted in official circles. Far from suggesting that there is any immediate need for the imposition of any kind of controls on the lake fisheries, one gains the impression that every encouragement is given for the fishermen to keep on stepping up output. In fact the whole trend in the use of the natural resources of the countries surrounding Lake Chad, including fish, forest products and wildlife, is at present strongly suggestive of a runaway horse.

Markets have existed throughout northern and western Africa for centuries, and a very sophisticated form of trade and commerce carried on, characterised by bargaining practised according to clear-cut rules. This bargaining is, of course, a serious affair, but it is also treated like a form of sport. The vendor first settles in his mind his own desired price to be obtained for his product. This is a bit higher than the known current market value. He demands a sum about $1/3$ again higher than his desired price. The buyer offers about $1/3$ of the price asked, at which the vendor throws out his hands in a gesture of refusal, invokes the name of Allah, and may even walk away, spitting expressively as he turns to go. After a while he casually reappears, and the buyer, equally casually, mentions a new sum, about 50 per cent of the first demand price. The vendor offers a small concession. The game warms up and the buyer improves his offer. The vendor concedes another fraction and, finally, a price is accepted, usually about 80-90 per cent of the vendor's intended price, or, in practice, the current market rate.

The game is time consuming but socially important. Sometimes Europeans and Westerners who are newcomers to these parts of Africa react badly towards it, feeling that it is in some way immoral or dishonest. However, if the game is played according to the rules, it is seen to be an authentic and effective approach to business.

The markets have always been the centres of trade and commerce, and *foci* for the intermingling of the tribes and the dissemination of information and religion. Among the more famous are Timbuktu, Kano and Maiduguri. Usually these markets sprang up along the ancient and well-established caravan routes, and in them imported goods from the far north were bartered for inland products, including slaves and ivory.

Numerous markets are scattered all around Lake Chad, in the major towns and villages. Among these Bol, Nguigmi, Bisigana, Baga, Kukawa, Mongonu, Marte and Wulgo may be mentioned for their historic markets, but most of these today present but a shadow of their former glory.

It is traditional in the Kanem and Bornu markets for all the dealers in similar or related commodities to congregate in their own special section of the market. Thus all the dealers in natron, saddlery, poultry, rope, or hardware (such as swords and daggers, hoes and machetes), may be located within their own group and their own section. Many of the commodities sold and exchanged today are the same as those sold over the centuries. Camel and horse saddlery, water jars, daggers, woven baskets and paniers, for example, have changed little.

Many commodities have changed merely in style and fashion, such as the pattern woven into the grass hats, baskets, mats, and local cloth. Some items have simply been modernised - like the Kanuri version of the 'piggy bank', These are hollow spherical pottery balls with an opening, made formerly to receive savings in the form of cowry shells, but now (1970) as coins in shillings and pence, or francs.

Yet other items have been partially or totally replaced by manufactured imported articles. Among these may be cited the dyes used for cloth and weaving grasses. Although balls of local indigo may still be purchased in just the same form as those described by Denham in his 1823 visit to Kukawa market, the same dealers today also sell packets of dyes made in Manchester and Birmingham.

The dyeing of cloth is an age-old industry in most of the cities and towns of the southern Sahara, cloth being dipped into dye which has been mixed in deep clay pits or huge pots sunk into the ground. We saw the pot variety in 1955 on our visit to Wulgo, and they were still using only local dyes then. In Maiduguri today (1969/70), both local and imported dyes are used, the imported ones usually appearing to be rather more vivid in colour than local dyes when fresh, but fading more rapidly with exposure to the sun.

Similarly one may find that the imported kerosene storm (or hurricane) lamp, and lamps made from empty condensed milk tins, have generally replaced the traditional pottery oil lamp characterised by its deep orange, warm, flickering flame. Alongside the lamp sellers, other dealers sell 18-litre tins of kerosene as well as their locally made pots filled with groundnut oil. Yet other dealers sell empty kerosene and motor oil cans and belong to the tinker section, many of them being adept at soldering useful household articles from these and other tins.

Then there are the gourd, or calabash, sections where all shapes and sizes of gourds are sold for classified uses such as for storing milk, carrying grain, as ladles or as measures, while a neighbouring section may sell plastic bottles, plates, buckets, jars as an entirely modern line serving similar needs. Elsewhere one finds scrubbing brushes replacing locally grown loofahs; moulded plastic shoes replacing the sandals made from car tyres (which in turn had previously replaced hand-embroidered leather) for daily use; and detergent soap powders, patent medicines, European perfumes and hair grease, replacing native concoctions and the perfumes of Arabia. Torches, torch batteries and bulbs; milk of magnesia, tablets for headaches, chocolate powders for making hot drinks; cigarettes, plastic toy crocodiles and lions, pencils and ball-point pens: all these are to be found on the stalls or on trays hawked around on the heads of small boys.

As in the past there are stalls and head-trays where a drink of local guinea-corn wine (said to be non-alcoholic, although liable to ferment if kept too long), kola nuts (to be shared as the symbol of friendship) and ground nuts can be purchased. In the past, however, it was considered a disgrace to 'eat on the path' and snacks were not really approved. Today, however, snacks are sold in every market and include meat barbecued on thin sticks around a fire, freshly roasted fish, doughnut-like cakes fried in fat and eaten with sugar, roasted sweet potatoes, and even platefuls of boiled rice with chicken, fish stew, or soup.

Bottled soft drinks and beer are generally readily available, although alcohol is strictly taboo for devout Muslims. Then there is usually a bookstall, frequently to be located on the head of a hawker. It usually contains primarily Islamic literature in arabic script, but nowadays it may also have the modern touch too, and contain associated items of stationery such as envelopes and pads of ruled writing paper, biros and pencils, exercise books and even some simple school books in English and Hausa.

The one and only kind of stall which I have never found to be modernised or superseded is the charm stall: here the dealers in charms connected with the occult trade their wares and their wisdom, offering, as from time immemorial, claws of civets, ostrich tendons, bits of skins of various wild animals, pieces of skin from an elephant's trunk, little horns stuffed with queer mixtures of magical substances, mini-portions of the Koran minutely inscribed and stitched into tiny leather pouches, and phylacteries for wearing on thongs around the neck or waist.

There are always sections in each market dealing in the various means of transport: in the past these dealt mainly with donkeys, horses, oxen and camels, but nowadays the trade is enhanced by the presence of bicycle hirers, sellers and repairers; motor mechanics equipped with gas cylinders and welding apparatus; tyre menders with ingenious home-made vulcanising tools; and the sellers of salvaged bits of vehicles such as exhaust pipes, nuts and bolts, sump covers and distributor heads.

In the markets, even those in remote places, there is a dramatic confrontation of the commodities of the modern technological world of effortless speed, with those of the ancient world of craft, manual labour and physical effort. Not only have modern manufactured articles intruded into these markets, but an African version of the travelling salesman appears. Where there is a motor road, he arrives in the polished company car, with jolly advertisements all over it, and public address system loud speakers on the roof. He steps out, looking cool and educated in his neat tropical dacron suit and sun-glasses, while his equally smart assistant goes to work over the loud speakers. Soon a crowd surrounds him, and in no time he is carrying on a roaring wholesale trade with the stall-holders.

Other businessmen offer various services to the public: building contractors offer better houses of cement block with corrugated iron roofs and concrete floors; taxi drivers offer a quick (expensive) direct service to the nearest city; transport contractors handle articulated lorries capable of carrying over thirty tons of fish, grain or cattle; tailors offer to make the mosquito net, the trousers, or the robe of your choice; and barbers offer the latest in men's hair styles, which may be selected from a board on which between ten and twelve styles are illustrated. Naturally these services are primarily offered to the 'modern' man, the man with an eye to the big chance.

I found it astonishing to see how wide is the range between wealth and poverty in these market towns and in the people who gather in them. The villagers from far away across the dunes come in for market day on foot, driving their patient little donkeys with burdens of grain, firewood, thatching grass or home-made mats for sale and exchange, returning in the afternoon or evening with some fish, perhaps a tin of kerosene, a couple of bottles of groundnut oil, some cloth, a new bridle for the horse, and a plastic bucket. Others may ride in on horseback, leaving their children to drive the donkeys. Yet others, from further away, may come in by camel, leading a train laden with goods grown in the north or traded in more northerly markets, or maybe with great blocks of natron. *(See: Plates 21. 2-4; 22.3-4)* These return with goods which are unobtainable further north, like iron four-poster bedsteads (very popular, even to the brass knobs, as they are suitable for hanging up a mosquito net) and perhaps inner-spring mattresses for the rich, tables, chairs, buckets, bicycles, nylon nets, enamel and plastic tableware, and kerosene.

And then there are the really poor: the aged, the blind, the crippled, and the lepers. In Muslim law these must receive alms from the faithful, and so they beg, their identity made known by the begging bowl or half-calabash in their hand. In the markets around the lake these receive little, but in practice most of them do not entirely depend upon alms as they usually belong to a household of which the head is bound to maintain them. Begging is thus done partly to compensate for their lack of practical contribution in the compound, and partly to provide a visible outlet for the rich to demonstrate their piety and generosity. Lepers are not outcasts and move freely among the populace, advanced cases sometimes being given a donkey on which to ride while begging. *(See Plate 32.5)* These lepers are a pathetic sight, often lacking nose, fingers and toes, and yet they are rarely overtly badly treated by their own people.

Even mentally ill people are not ill-treated as long as they are not dangerous, and I have frequently seen a woman at a snack stall, or a meat vendor, fill the begging bowl of a mentally ill person, cripple, blind person or leper, with food. Very aged people have few comforts, and yet in this type of society they too usually belong to a large extended family and would be cared for there. Sometimes, nevertheless, the lot of the aged is hard.

Poverty, ignorance and disease are frequently named in developing African countries as enemies to be overcome, as indeed they are. Yet, because of the closely integrated family and household system still operating in these remote areas, even the poor can survive and few are left alone to fend for themselves as are the poor in so many of the so-called civilised countries, and in the big cities.

These days (1970) the main markets are connected by motor roads with Fort Lamy (N'Djamena) and Maiduguri, but the bulk of the rural people still depend on their animals for transport. With the Kanembu-Kanuri, the Shuwas, and the Fulani, horses and camels retain pride of place for transport and ceremonial. Many of the wealthier Kanembu-Kanuri, and Shuwas, travel on horseback to important functions such as council, business and social engagements, the horses decked in elaborately decorated saddlery and their riders in expensive flowing robes, usually attended by servants, either on horseback also, or on foot. For longer distances, especially to the north of the lake, camels may be ridden. In some cases, however, four-wheel-drive vehicles, and even luxury 4 x 4 limousines, have replaced horses and camels for such occasions.

Less wealthy peasants may also use horses or camels, but the horse's saddlery and rider's robes are simpler, and where horses are used in the day's work for drawing water or herding cattle, the saddle may be replaced by panniers for water pots or kerosene tins, or be absent altogether; and the bridle may be replaced by a single rope lightly tied around the lower jaw, or a simple head-collar.

Similarly, camel saddlery is directly related to the rider's status and purpose, but in its most basic form the bridle simply consists of a rope around the lower jaw, and the saddle a piece of leather.

Every rider takes with him a set of foreleg hobbles, as hobbling is the universal method of restricting all domestic animals from fowls to camels (with the exception of dogs and cats) around Lake Chad. On arrival at the rider's destination, the animal is unsaddled, hobbled and let loose *(See: Plate 22.1)*. A camel may be left kneeling with a band around one knee which prevents it from rising, and a horse may in addition be tethered by a hobble line from one forefoot to a mushroom-shaped peg in the ground. It is customary for the Kanembu-Kanuri to leave their horses and camels out of doors at all times, although some chiefs may have covered stalls within their compounds. The Shuwas, however, who hold their horses in especial esteem, customarily provide them with shelter. The Fulani keep horses purely for utilitarian reasons, rarely using elaborate saddlery on them or giving them special attention, but, for them, it is the cattle that hold pride of place.

Kanembu-Kanuri women take part in the care and handling of all livestock, as do the men, and they also ride horses, sitting astride, and handling them competently and confidently. Horses are allowed to breed freely around Lake Chad, and mares as well as stallions are used for riding. In contrast, in the Hausa states, mares may be kept in the compounds but are rarely used for riding. Mules are very rare indeed around Lake Chad. Three main breed lines are recognisable in the horses around the Lake: the small, tough barbary pony, the nubian and the true arabian. From time to time the French have imported French thoroughbreds to Tchad and Niger, and occasionally horses are to be seen around the lake with features indicative of this influence. The typical 'bornu' horse is a cross-breed incorporating all three breeds in its ancestry. Characteristically it has a markedly 'roman' shaped face as seen in profile, stands between 13 and 15 hands high, and comes in predominantly roan, roan mixtures, piebald, skewbald, sorrel, bay or grey colours. Occasional true whites occur, with pink skin underlying white hair. These whites are much favoured as the riding mounts of chiefs. They tend to suffer acutely from sunburn, and are sometimes accorded the privilege of being sheltered, even by the Kanembu-Kanuri.

Barbary ponies rarely exceed 14 hands in height, and are frequently piebalds and skewbalds. Horses bred in the Bahr-el-Ghazal are generally larger and show a strong nubian-arabian influence. These may come in 'desert-mix' colours, and occasionally as true whites, but blacks and very dark bays also occur. True arabians are bred in the Bahr-el-Ghazal and are of course distinguishable primarily by the dished head profile, grey colour range, and supple build *(See: Plate 35.5 and 6)*.

## Chapter 9 - Fish, livestock and markets

So much of the history of the Sefuwa dynasties is bound up with horses that it is not at all surprising to find that much of the saddlery, trappings, and armour displayed during traditional ceremonies consists of ancient treasured items handed down within the royal households. Where the originals have disintegrated and become unserviceable, replacements with a modern flavour may sometimes be seen on high days and holidays. The rider with the true ancestral panoply may be wearing embroidered war robes, embroidered camel-leather boots, silver-braided sash, ornamented helmet, shield, sword and buckler, lance and pennant, and genuine, if torn, chain mail.

The horse may have an ornate bridle with engraved silver pieces, colourful tassels and pompoms, an elaborate and vicious bit, a silver-covered breastband laden with fringes of silver-threaded tassels, a high pommelled saddle with silver pommel hoods bearing the Islamic star and crescent, elaborate and vicious stirrups, and numerous saddle cloths above and below the saddle, some threaded with real silver and gold thread. It is not unusual also to see the Christian cross, in a sort of 'Ethiopian' style, on some saddle pommels. In some cases genuine, ancient armour plates cover the horse's head and neck, while the body is dressed overall with a quilted blanket reaching to the hocks. A modern variant of the latter may be a pyjama-like garment of satin or cotton with a floral pattern worn like trousers on the horse's forelegs, and reaching to the fetlocks, while at the rear it spreads almost to the ground over the hindquarters and tail, like a royal train.

Other traditional elements to be seen on such occasions are mounted drummers, pipers, and trumpeters, as well as the ever-present court jester. *(See: Plate 22.5)*. At Kukawa, the jester is traditionally dressed in a black robe with the mounted head of a ground hornbill *(Bucorvus abyssinicus)* as a crest above his black hood. The ground hornbill, a black and white bird about the size of a turkey, was commonly found throughout the northern savannahs, and, as it frequently associates with wild animals, a hunter would mount a hornbill's head on his own head and then stalk through the grass towards his quarry, nodding the mounted head to simulate the live bird. In this way he hoped to make a very close approach and ensure a successful shot. I did not discover the significance of the Kukawa jester's rigout, but presume that it signified stalking success in hunting and possibly also in war *(Plate 8.1)*.

Camels also take their place in traditional ceremonies, and again they are decked in ancestral saddlery and emblems. The cameleers at such times love to demonstrate their skill, galloping and running the camels at great speed, making the camels do obeisance, walking on their knees or kneeling while their armed riders mime

## Chapter 9 - Fish, livestock and markets

the laying of dune-land ambushes, and then shooting across their humps at imaginary advancing armies.

Camels are useful and efficient, if slow, as transport animals around the lake, for they can not only cope with sand, but also with the mud of the lagoons and *firki* pans. Tuareg and Arab camel caravans still regularly cross the desert to and from North Africa bringing embroidered cloth and robes from as far away as Damascus, and perfumes, jewels and other treasures from Egypt, Libya, Algeria and Morocco. Many Kanembu-Kanuri households own one or more camels, and some of the Bornu Fulani use them a great deal.

Being accustomed to ranch-type horse riding, I had no difficulty at all in accommodating myself immediately to the motion of a camel, although some people find it very tiring at first. Camels are by no means gentle beasts by nature and can be extremely obstinate if not handled properly. They also have a very disconcerting way of turning their heads round when kneeling baring their teeth and growling in one's face as one mounts or dismounts. Fulani cameleers frequently mount them with great agility by running up the tail or taking a step up via the neck of the standing animal. The camels to be seen around Lake Chad are the arabian or single-humped type, usually sandy or fawn in colour, although both white and variegated ones also occur. In 'freight' camels, the hump sometimes becomes very misshapen due to the continual pressure of the pack saddle, as well as its depletion as a food reserve on long journeys across the desert.

In normal circumstances a camel never rears like a horse, and it is very rare indeed for one to fall. However, during the Greater Beiram Festival celebrations at Kukawa in February 1970, I did see a running camel trip head-over-heels over a little donkey that strayed across its path, and fall headlong. Neither camel nor rider (nor the donkey) was hurt and the *status quo* was speedily restored.

Camels do occasionally bolt, and a Nigerian missionary of the African Missionary Society was thrown and left by his camel one day when he was travelling through the lagoons at the north end of the lake. He suddenly found they were face to face with a startled elephant. The missionary fell into the shallow waters of the lagoon so near to the elephant that he recalled that when the elephant's ears flapped, his shirt also flapped. The villagers, seeing the riderless camel come rushing home, hastened to the place to see what had happened and saw the elephant making menacing gestures. Apparently their shouts disturbed it and it went away leaving the very damp, frightened, but unharmed, missionary in the water.

The Fulani, like the Shuwas, are essentially pastoral by nature and inclination. Cattle are the most important things in their lives. So beloved are their cattle that a beautiful woman may be described in terms such as: 'she has the eyes of a heifer and the breasts of a milch cow'. Formerly their herds consisted exclusively of the large white fulani breed, but nowadays (1970) they also contain large numbers of red bornu and the heavy-horned Lake Chad *(kourri)* cattle. Unlike the Yedina, the Fulani people use curdled milk and cheese in preference to fresh milk, although they do also drink both goat's and cow's milk fresh on occasion. The Shuwas also have a curious custom: when their pack and riding oxen return to camp after a long day's work, they bleed them from the nostrils, claiming that this treatment improves performance.

Both the Kanembu-Kanuri and the Fulani invariably also keep large flocks of sheep and goats, as well as numbers of donkeys. The sheep and goats are rather tall and gangly, resembling each other in general size and appearance, coming in various patterns with spots, stripes and blobs. The sheep have long, dangling ears and tails, while the ears of the goats hang laterally away from the head and the tails are upright. I made several sample counts to estimate colour pattern percentages in these herds and found a very fair 3:1 Mendelian inheritance ratio of pigmented sheep and goats to pure whites.

*Lake Chad versus The Sahara Desert*

# PLATE 29
# JANUARY 2002 (C)

1. *(above left).* The newly built Lycée. 2. *(above right).* The former secondary school.

3. *(above left).* Primary school children and their school building.
4. *(above right).* A village church on the route from N'Djamena to Bol.

5 and 6. *(above).* Dressed in their 'Sunday best', members of the Evangelical Church are seen here after Sunday worship: men and women are seated on separate sides inside the church, and leave separately at the end. 7. *(below).* Heavy, powered freight canoes.

# PLATE 30
# January 2002 (D)

1. *(right)*. Mosque (right of picture) on a dune island near Bol.

2. *(above left)*. The Imam of the mosque plaiting grass rope.
3. *(above right)*. Kourri cattle in Bol area *(May 2000, photo Haberkamp)*

5. *(above right)*. The waterfront in Bol: a busy trading trading port for goods transported across the lake between Bol, Baga, and N'Djamena.

4. *(above left)*. The butcher in Bol.

# PLATE 31
# JANUARY 2002 (E)

1. *(above left).* 'Development' includes oil production. This oil pipe line has long carried oil across the Sahel from Rigrig southeastwards to a refinery in Tchad. A massive oil extraction and export project to Cameroun is in progress. 2. *(above centre).* At Bol lake levels are measured regularly. The old gauge is high and dry due to lake regression; a new one is seen in the lower level water beyond.
3. *(above right).* Irrigation channels in the Bol polder.

4. *(above left).* Due to the regression of the lake, papyrus has dried up and is not sufficient, or long enough, for kadai-building: an enterprising fisherman has used corn stalks instead, but they do not last as long or float as well. 5. *(above right).* A fisherman sets his net.

6. *(left)* The local ferry.

7. *(right)* A very small fish trap.

# PLATE 32
# CHRISTIAN MISSIONARY WORK

1. *(left).* The float plane of the Mission Aviation Fellowship in use during the 1969/70 expedition: the pilot, (seen here with children and helpers) rescued us when the engine of the *Jolly Hippo* yacht failed.

2. (*above left*). In 1974, children on an island had made their own model of the MAF plane then in use on the lake, using dry papyrus fronds.
3. *(above right)* The 'birth' of the new Mission Station at Haikalu island in 1969. It was sadly destroyed by rebels c.1978.

4. *(above left).* Missionary pastor and the church and congregation on Kinjeria island in 1974. 5. *(above right).* Health care is very limited on and around Lake Chad: a man with leprosy, no hands or toes, is dependent on his precious donkey as he begs for food and clothing. Government and Mission clinics are few and far between.

## PLATE 33
## TOWN CONTRASTS

1. *(left).* Baga town, 1969/70, a town of grass and mud brick houses in grass-walled compounds on the sandy ridge of the Baga peninsular. By 1974, some cement block buildings had appeared, some built below the lake's inundation levels: the lake has not yet returned again to flood them.

2. *(above left).* The great Chari River with fruit vendors on its shore.
3. *(above right).* An ivory seller in N'Djamena (1957). His wares represented the significant number of elephants killed then.

4. *(above left).* A building in Fort Lamy (1957).
5. *(above right).* After Sunday worship in one of the N'Djamena Evangelical Churches (2002).

# PLATE 34
# HERE & THERE AROUND LAKE CHAD

1. *(above left).* The Chari River (January 2002) seriously reduced primarily due to diversionary uses and reduced rainfall, especially on its watersheds with the Benue and Oubangui rivers (Cameroun top left; N'Djamena bottom right of picture).
2. *(above right).* Cast-net fishing at dawn on the Chari (1957).

3. *(above left).* Plank canoe on the Chari river (1957) using an old style of block-ended paddle reminiscent of the ancient Yedina paddles.
4. *(above right).* Fisheries research vessel on Lake Chad.

5. *(above left).* The forge at Bol (January 2002) making anything from daggers and axes, to fishing spears and bits for horses' bridles. The smith (right) forges the iron while his assistant (left) works the leather bellows for the fire.
6. *(above right).* Soil contaminated by natron deposits from an earlier lake overflow lies dry, exposed and useless for cultivation.

*Lake Chad versus The Sahara Desert*

# PLATE 35
# LAKE CHAD: MAINLAND SAVANNAH (A)

1 and 2. *(left).* A pair for wild West African crowned cranes and *(right)* a crowned crane chick kept by a Kanuri fishing family on Lake Chad as a pet.

3. *(left).* African fish eagle on Lake Chad. Many times we heard its cry ring out over the lake and saw it swoop down to seize a huge fish and then rise again with it in its talons.

4. *(right).* A ratel (honey badger), rarely seen, but still present on the southern and eastern mainland around Lake Chad.

5. *(above left).* A village chief on his sturdy 'Bornu' horse, a cross-breed probably incorporating nubian and barbary *(see page 285).*

6. *(above right).* The author's Bahr-el-Ghazal polo pony: an arab-french-thoroughbred cross, formerly used for racing.

# PLATE 36
## LAKE CHAD: MAINLAND SAVANNAH (B)

1. *(above left).* A line of baobab trees (near the town of Kukawa) possibly seeded long ago from the droppings of migrating elephants.

2. *(above right).* A huge bull elephant dominates the scene.

3. *(above left).* Young bulls are often playful.   4. *(above right).* West African giraffes, formerly found all round Lake Chad, are no longer found on the southern and western mainland, but are protected in Cameroun to the east.

5. *(above left).* Lesser bustard.

6. *(above right).* Vultures: always on the watch for food.

Jacob* clearly knew just what he was up to when he manipulated the breeding of his uncle Laban's flocks to his own advantage. His animals were undoubtedly very similar to those found around Lake Chad today, and indeed these may be descendants of his flocks.

The donkeys of this area are usually very small, strong, and grey in colour, although whites and blacks also occur. They are never saddled or bridled (except in the case of having a rider with leprosy), being directed entirely by movements of a short stick over the head and neck. Padded sacks are sometimes used under their loads, but they often suffer from terrible back sores. They are frequently ridden bareback by men, women and children.

Every Kanuri-Kanembu compound has its quota of poultry consisting of scrawny, athletic-looking fowls, muscovy ducks, guinea-fowl, and aberrant guinea-fowl which are thought by some to be the result of crosses between domestic fowl and guinea-fowl.

Although in many countries dogs are incompatible with the religion of Islam, they are frequently kept by Muslims in Kanem and Bornu, and by the Fulani. Every Fulani camp has its quota of dogs, most of which belong to the most basic of all breeds, the original 'dog' (basenji) of Africa and the Mediterranean lands. This is a lithe creature, usually sandy, fawn or brindled with triangular

---

* The Holy Bible KJV Genesis. 30 v 39

ears standing out from the head, amber eyes and a white tip to the tail. The bitches like to whelp in burrows or sandy hollows if possible. These dogs are intensely loyal in spite of many privations and constantly associate with the herds and herdsmen. The Fulani cut off the tips of their ears, which are very vulnerable to the ravages of the *Stomoxys* fly.

A typical Fulani camp in the vicinity of Lake Chad consists of temporary tents made of mats hung on poles or from acacia trees. The women and girls live in the camps and travel with the men at all times. Unlike the Yedina, the Fulani women also take part in the herding and milking, and are also responsible for the cheese making. Calves are usually tethered at intervals by one foot to a rope run along the ground, and certain individual leader cows and bulls may be hobbled at night, while the main herds cluster around them.

During the day the herds are led to and from the water holes by Fulani men on foot. The herdsman carries a slender staff at all times. While the herds graze or drink, he uses it as a leaning post, standing on one leg with the other crooked against the opposite knee, but when it is time for the herds to move on, he places the staff across his shoulders, dangling his extended arms by the wrists over it, and walks off in the chosen direction. This is the signal for the herds to follow, and wherever he leads in this way, there they also go. Sometimes he whistles up a straggler, or carries a very small calf, lamb or kid on his shoulders, and sometimes a sick calf is loaded on to the back of its own mother or on to that of one of the pack oxen.

In the past the clay pans and inundation areas provided water for part of the year for the Fulani herds, and when these sources of supply dried up, they migrated southwards in search of greener pastures and permanent rivers, returning the following rainy season to the shores of the lake. On these migrations the pack animals were loaded up with the household utensils, camp equipment, smaller children, and calves, kids and lambs. These were mainly driven by women while the men went ahead with the main herds, pasturing them along the way as opportunity offered. But with the opening up of artesian wells every 15kms or so apart, all this has changed in northern Bornu and some parts of Kanem, and today (1970) the Fulani and their great herds cluster around these pools throughout the dry seasons, fanning out each wet season into the old areas around the flooded *firki* pans and inundation zones *(See: Plate 23.1-3).*

Nowadays (1970), as each dry season wears on, every blade of fodder, every twig and root around the artesian overspill pools is eaten, the bare sand extending daily as an ever-widening circle, until not a single edible straw remains within a radius of anything up to three or even four kilometres.

## Chapter 9 - Fish, livestock and markets

The animals get thinner and thinner, and each day several lie down near the pools and never get up again. *(See: Plate 23.4)* The vultures gather around, kites circle and marabous hesitate nearby, all awaiting the moment when they can move in and strip every shred of flesh from the bones.

In the driest part of the year - about March and April - flocks and herds numbering hundreds may be seen at intervals throughout the day approaching each water hole. As the thirsty animals draw near, they break into a run, plunging deep into the precious water in their combined desperation to drink and to stave off the hunger that has now become their constant companion. Horses are usually privileged by having water drawn from the well mouth itself and handed to them in half-calabashes, but camels must drink direct from the pool. They approach the crowded brink with their usual air of apparent disdain and unwillingness, and sip the water haughtily as if embarrassed by their own public admission to having, even if not every day, to drink water like other animals.

Women assist weaker and younger animals to drink by drawing water for them. I watched a camel which had given birth to a calf in the night: she was saddled and had been ridden to the pool, her gangling baby at foot, but there at the pool the Fulani women had allowed her to remain kneeling while they drew innumerable calabashfuls of water for her.

Hand-watering a long train of camels is a wearying, time-consuming business, and I have often paused to ponder with amusement the aptness of the test chosen by Abraham's servant when he was sent with a train of ten camels from Canaan to Nahor, in Mesopotamia, to find a bride for Isaac. The test he chose was to ask the girls at the well for a drink. The one who should say she would draw for him and for the ten camels would be the bride for Isaac. Rebekah turned out to be the one who did this and became Isaac's bride.*

Thus the life not only of all the mainland tribes, but of the Yedina as well, is essentially bound up with the life of the livestock which comprise their wealth, means of transport, and source of food.

Formerly mainland cattle were not slaughtered for food unless weakly or sick, the number of head owned by any one household being far more important than quality. Cattle were a kind of currency, while horses and camels - also a sign of wealth - were primarily status symbols. Donkeys were, and still are, the peasant's beast of burden and the poor man's main (and often only) means of transport. Sheep, goats and poultry, as in the past, provide meat for food and serve as sacrificial animals.

---

* The Holy Bible Genesis. chapter 24.

Many of the more progressive of the Kanembu-Kanuri and Fulani trade their livestock for cash, but usually sell off only the poorer quality animals. The majority are still strongly resistant to the idea of marketing high quality beasts or managing smaller herds with a view to improving quality rather than quantity. Although horses and donkeys are never used either as food or for sacrifice around Lake Chad, old and decrepit animals are sometimes sold to traders from coastal areas to the south, where it is said that horses are used for ritual sacrificial ceremonies that have replaced a former, even more barbaric, system.

A more lucrative outlet for horse-dealing, however, although its market is limited, is the sale of high quality horses for racing and polo. Bahr-el-Ghazal horses are especially favoured for racing, while the bornu breed makes a very fine polo pony. Racing and polo are popular both with the wealthier indigenous people and with foreigners in Tchad and Nigeria, the Nigerian élite yielding some first-class polo players.

The rapidly increasing momentum of economic growth and development on and around Lake Chad is clearly manifest today, and whatever controls and regulatory measures may be imposed by future progressive governments, it is apparent that, no matter what may happen to the poor, the rich will become richer and richer.

But the mainland markets will lose none of their vitality or fascination, for, with whatever impetus modern trends may intrude and revolutionise, they will long continue, as in the past, to be the essential tribal, commercial, political and religious meeting places for native and foreigner, rich and poor, ancient and modern.

On the lake too, the fishermen will continue for as long as possible to pole and paddle their canoes, camp on the insecure and undulating floating islands, and net as many fish as necessary to satisfy the needs of the commercial 'sharks' on the mainland.

*Across the dunes, the livestock also must long continue to play a vital role in the life of the people. The cheerful horseman with his sword on his thigh, his conical straw hat, and his flowing robes will yet be seen cantering along the sandy paths; the patient little donkeys with their enormous burdens must still continue to stagger faithfully under the burning sun to and from the overcrowded waterholes and seek out those precious token patches of shade; while the hooded and turbaned cameleers, on their stately, eternally supercilious camels, will continue to stalk across the dunes, their eyes ever focused on the distant and hazy horizons of wisdom and wealth.*

# Chapter 10

# *Modernisation and enlightenment*

*Better* farming, *better* animal husbandry, *better* transport, *better* irrigation, *better* government, *better* fishing techniques, *better* boats, *better* standards of living, *better* health, *better* schooling . . . and a *better* life: technology is the accepted modern means of providing material improvement in so-called 'developing countries', and every African government today strives by all available means to compete in this progress race.

But new techniques need new tools, and new tools need trained and reliable operators. Then technical programmes and personnel need coordination, and coordinators need not only knowledge of, but insight into, the ever-changing socio-political norms of communities which may be involved. They need not only to be able to assess trends and measure potential extremes, but also to communicate these effectively both to the rural communities and also to the representatives of highly volatile governments who may, or may not, take appropriate action. And, finally, programmed technology needs to be fed back to the communities concerned in terms that can be understood and believed.

The past three decades (1940s to 60s) have seen a wide range of developmental programmes set up and put into operation within the Chad basin. Many of these have been achieved by specific departments of the governments of the four nations involved, some through private enterprise and some by international agencies. Among these may be cited the building of the major trunk roads from Maiduguri to Baga on the lake shore, from Maiduguri to Kano, and Maiduguri to Bama.

The research programme that led to the development of the artesian system of the Chad basin for industrial and community use is another. The development of pilot schemes for modern irrigation and large-scale wheat growing on the western and southern sides of the lake, and the development of polder farming in Kanem are also major technical enterprises. The introduction to the lake itself of nylon fishing gear, motorised vessels, amphibian aircraft, and hovercraft, also suggests wide developmental possibilities.

Two major problems, however, emerge as one tries to grasp the real perspectives of these developments on and around Lake Chad. The first is the question as to how viable modern technical programmes really are when they are almost entirely dependent upon the complex tools and machines of modern technology, and upon operators highly trained in specific technical skills. The second is the question as to how safe these programmes are if operated in isolation from each other within a complex, and very fragile ecosystem, of which the natural regulatory mechanisms are still largely unknown.

The first of these problems is sociological. Faced with the ever recurring fact that the majority of Africans do not characteristically possess the sense of exploration and curiosity that leads either to a career in scientific research, or to mechanical inventiveness, it is logical to question the wisdom of setting up large-scale schemes dependent on the successful use and maintenance of modern machinery in remote areas.

Annoying as indigenous governments have found this to be, it has (up to now - 1969/70) been necessary repeatedly to employ contracted technicians and research officers from elsewhere, not only to install, but also to maintain and supervise such schemes. No matter how vigorously indigenous technical trainees are encouraged and provided for, they rarely manage to filter back permanently to remote rural areas to man such schemes on a sustained basis. There are other opportunities with more attractive features that do not make such heavy demands on adaptability, inventiveness, foresight, drive, individual responsibility and the willingness to make decisions personally.

Considered in these terms, one questions, for example, the wisdom of setting up large-scale wheat schemes where the harvest is entirely dependent upon a sustained water supply pumped into permanent raised channels across a stretch of semi-desert. When the pumps fail one by one, the mechanics go off sick, the spare parts do not arrive, the supervisor is on leave and the senior research officer is neither willing nor able to take the pumps to pieces himself, there is a reasonable possibility that the entire harvest may fail to mature, let alone reach its market. There may be no other qualified people to be found within a radius of 300kms, even if they are willing or could spare the time to help out.

In any case, in a land where addiction to corruption and bribery are the most dangerous diseases known, financial safeguards are so rigid that it would not be possible to mobilise funds to employ anyone at the necessary level of qualification on a temporary basis. To attempt to run such schemes at a sustained profit is therefore not only unrealistic but liable to prove self-destructive: that is, unless it is an accepted fact from the start that it must be manned by a contracted technical staff.

When we compared the wheat scheme with a market garden-cum-farm, run and developed by an enterprising Kanuri family in an inundation area of about one square kilometre near Baga, it was readily apparent that this family was making a sustained profit, with virtually no capital outlay or overheads, and no complex modern machinery or permanent structures. The ground was tilled by hand, but the farmer hoped soon to buy a yoke of working oxen and a plough. He irrigated the land in the dry weather by means of an old fashioned shaduf from a shallow groundwater well, and in the inundation and rainy periods channelled the water between ridges *(See: Plate 23.5 and 6).*

The *shaduf i*s a pole, weighted at one end with clay, and pivoted on a strong frame made from local acacia. To the other end, the longer arm, a rope carrying a calabash is attached. It is a simple, effortless movement to drop the calabash into the well, raise it, and swing it round on the shaduf so it can be emptied straight into the head of the irrigation channel. No tractor is needed to build the irrigation channels as they are simply raised by means of locally made hand-hoes. No permanent structures are involved and whether the lake rises or falls no major damage can accrue: the farmer then simply moves nearer to, or further from, the lake according to its mood.

As this farmer accumulates profits he may use them to invest in more wives and concubines - that is, an increasing potential labour force. Alternatively he may invest in more modern equipment or join a co-operative society whence he can obtain the occasional services of a tractor and driver (without depending on them), supplies of fertiliser at reduced cost, and even the loan of a pump. He may even begin to add new crops to his farm such as wheat or rice if the soil proves suitable. His economic success will result from the fact that he is independent of the use of complex technical instruments and their operators, an imposed business hierarchy, the installation of permanent physical structures, and the need to operate on a scale, and in a manner, beyond his own capacity. Above all, the enterprise is his own, and requires no training beyond what he can obtain locally in rural advancement classes.

The second major problem relates to the question of the political and technical coordination that is essential, if the risk inherent in the operation of large-scale technical developments around Lake Chad is to be contained. Any large-scale operation, developed in isolation, which may affect the hydrological balance of the lake, could possibly bring disaster to the entire area. To meet this need for international and technical cooperation in planning and control, the Chad Basin Commission was set up by the four participating governments.

As early as 1952, a resolution relating to the necessity for international cooperation in the regulation and use of water was adopted by the United Nations Organisation. In 1963 the Organisation for African Unity recognised the need to formulate principles for the utilisation of the resources of the Chad Basin for economic purposes, and the four member nations recognised the dangers that might result from the unrestricted exploitation of the water and other resources of the basin. Already they saw a rapid build-up of developmental pressure on the utilisation of the lake and its feeder rivers, the groundwater system, the other natural resources, and the people.

The Chad Basin Commission, comprising two Commissioners from each member nation, and a comprehensive administrative staff, was therefore established and its terms of reference defined by a Convention, and signed by the heads of the four participant nations, namely the Republics of Tchad, Niger, Nigeria and Cameroun. The following principles were stated:

> 'The development of the Chad basin, and in particular the utilisation of the surface and groundwater are given their widest connotation and refer in particular to domestic, industrial and agricultural development, and the collection of the products of its fauna and flora.'

> 'The member states must abstain from taking, without prior consultation with the Commission, any measures likely to have an appreciable effect either on the extent of the loss of water or on the nature of the annual hydrologic changes within the basin, the sanitary condition of the water, or the biological characteristics of its fauna and flora.'

'In particular each state must abstain from carrying out, on the portion of the basin under its jurisdiction, any hydraulic works or soil scheme likely to have an appreciable effect on the flow of surface or subterranean water in the basin without prior consultation with the Commission.'

Member states, however, were left free to implement already existing schemes and projects, providing that such schemes had no adverse effect on the Chad basin water regime. The following were then laid down as the primary functions of the Commission:

(a) to prepare general regulations which will permit the full application of the principles set forth in the Convention . . . and to ensure their effective application;

(b) to collect, evaluate, and disseminate information on proposals made by member states and to recommend plans for common projects and joint research programmes in the Chad basin;

(c) to maintain liaison between member states to ensure the most efficient use of the water of the basin;

(d) to follow the progress of the execution of surveys and works in the Chad basin as envisaged by the present Convention, and to keep the member states informed at least once a year thereon, through systematic and periodic reports which each state shall submit to it;

(e) to draw up common rules regarding navigation and transport;

(f) to draw up staff regulations and ensure their application;

(g) to examine complaints and promote the settlement of disputes and the resolution of differences;

(h) generally to supervise and implement the provisions of the present Statute and the Convention.

The Commission thus came into being as a high-powered, politico-diplomatic organisation, carrying weighty responsibility. Its establishment was a wise, far-sighted and vitally necessary move. Within its prescribed terms of reference, it has, during the first seven years of its existence, begun to function actively, but its real effectiveness can only be assessed over a much longer period, and it remains to be seen how well it will manage to communicate its developmental and coordinating function at all levels.

While this august body has ready access to scientific and economic data, technical advisers, and all the funds and instruments necessary for the application of desirable and necessary techniques, it unfortunately (like all politico-diplomatic bodies) lacks one very important facet. It lacks the means to harness human goodwill and effort in unified positive endeavour, while simultaneously applying restraints and economic discipline. This is because it lacks the means to overcome the human selfishness and greed for status and wealth that refuse to recognise the need for restraint in the use of natural resources, and because it cannot deal with the moral and social issues that cause inter-person, inter-tribal and inter-national incompatibility.

It is therefore inevitable that unless the Commission can communicate its decisions in a manner resulting in practical, unified endeavour and willing restraint on the part of the indigenous people, its efforts and indeed its very existence must prove barren.

Lines of educative communication around Lake Chad are still (1969/70) very limited. A school system exists, but the schools are few and far between and only the privileged minority ever attend even a primary school. There the few learn to read and write, to do simple arithmetic, and perhaps to learn a little theoretical geography and hygiene *(See Plate 19.2 and 3)*. A handful of school leavers are selected for secondary school entrance and another handful for teacher training. An even smaller handful may qualify for entrance to trade or technical schools, while the majority return to join the family business of farming, trading or fishing.*

---

* Educational facilities have improved since 1970: see ch.12

As so few can read and write, the dissemination of educative books and pamphlets is of limited value. Much more effective as a means of communicating with the Lake Chad public is the use of mobile cinemas, informative talks over public address systems mounted on vehicles, and public demonstrations of techniques. The problem with these media to date has been the lack of suitable material for communication, and the lack of suitably briefed exponents.

Of all the means of communication available on and around Lake Chad, by far the most effective and far-reaching is the radio, and without any doubt at all the radio speaks there with the most powerful of all voices. However, the radio, like the press, has many voices. On it the voices of the participant governments come over loud and clear in French, English, Hausa, Arabic, and Kanuri, carrying a mixture of educative information along with, and sometimes not clearly distinguishable from, political propaganda.

Over the radio also come the voices of the great business houses (advertising their varied products irrespective of the environment of application and use), of the world of entertainment (with its diet of sex, violence, crime, laughter and rhythm), and of varied political ideologies and religions.

In every village household, every market place, every floating camp, many canoes, and most camel trains, these voices crowd one another over transistor radios. To which voices do the people, the proud Kanembu-Kanuris with their centuries of Islamic teaching, and literacy; the nomadic Fulani and Shuwas with their beloved cattle and horses; and the ancient, almost totally illiterate Yedina listen?

Sometimes, as we slipped past a floating camp or fisherman's canoe of an evening, we could hear the radio voices echoing across the darkening waters. Sometimes we could hear the flow of rapid French from Fort Lamy (N'Djamena), and sometimes the coarse rhythms of some western 'pop' group; sometimes we heard West African news broadcasts in various languages, and sometimes a programme in English from Moscow or even French from Peking. Sometimes we heard Muslim religious programmes in Hausa and sometimes Christian services and Bible readings. There is no doubt that the ubiquitous transistor radio brings its listeners on and around Lake Chad a varied and confusing diet.

As I twiddled the knobs of my own little radio aboard the *Jolly Hippo,* there was always one station that came over clear and strong, sometimes in Arabic, and sometimes in Hausa, French or English: this was Radio E.L.W.A., the Christian station, broadcasting from Monrovia, Liberia.

I also frequently encountered fishermen on the lake, and traders in the markets listening to this station. Perhaps the main reasons for its popularity on the lake were the religious character of the people, its clarity and its use of a variety of languages. Not only was its reception usually clearer than that of most other stations, but the careful diction of the speakers, in whatever language they spoke, contrasted strongly with the jumble of ill-articulated words pouring over many of the national stations.

The broadcasts of Radio E.L.W.A. had also recently become meaningful to at least a few of the fishermen on Lake Chad, for in 1960 Dr David Carling, of the Sudan United Mission, began to realise the need of the lake people both for medical help and for spiritual enlightenment. Up to that time no medical facilities existed on or near the lake, and, as explained earlier, the religious influence of Islam had only recently come to the Yedina people in the form of a useful religious umbrella under which they could shelter conveniently for the furtherance of productive and peaceful commercial engagements with the mainlanders.

The Christian Gospel with its integrating and stabilising effect on individual life in the present, and its promise of practical and useful eternal life in the future, was still (1969/70) apparently almost unknown on Lake Chad.

Islam is a religion of discipline and it has, to some extent, had a unifying effect on tribes which embrace it. In the development of its household hierarchy, it also has a stabilising and dignifying effect. Literacy as associated with Islam has tended in the past to be limited to the élite, but the discipline engendered through its religious exercises, and through the influence of its malams (teachers) and imams (priests) affects all its adherents, both rich and poor. Through its use of Arabic as its religious language, there is a tendency for Arabic words and word roots to permeate local languages where its influence is felt, and these tend to form the basis for communication through the development of a lingua franca.

In the religion of Islam, Allah (God) is remote, impersonal, and unknowable, and the concept of godly and neighbourly love in its highest, selfless and divine meaning is not recognised. Islam offers no release from bondage, no opening of blind eyes, no healing of the sick, no forgiveness for sin, no saving of the lost, and no raising of the dead. This is inevitable because Muhammad did not claim to be more than a prophet, and thus was subject, himself, to all degrees of human failure and inadequacy. He could not raise man higher than his own human level. So it is not at all surprising to find that many educated Muslims today recognise the inadequacy of this religion, although few actually turn to godless materialism in its place.

## Chapter 10 - Modernisation and enlightenment

Indeed, it is very rare to meet any African who does not believe in the existence of a supreme spiritual Being who created man and the universe, although some recognise His practical operations in the affairs of men only through the media of lesser spirits (good and bad) working through natural objects and phenomena. This is the essence of animism, as held by most of the indigenous tribes that were absorbed by the Sefuwa ruling houses in the past, and later converted to Islam. It is therefore quite usual to find modern Kanembu and Kanuri people practising a sort of combined animistic and Islamic faith. In the case of the Yedina the Islamic veneer is, however, extremely thin.

When Dr Carling made his first medical trips on the lake by boat, the cases for medical treatment were few and scattered - an injured foot which had become infected, an abdominal hernia, fever, and so on - but soon people began to hear the news of his healing work. Dr Carling is a very tall, well-built man, with warm, gentle eyes, and a kind friendly smile: features that appeal greatly to the Yedina people with their own fine physique and open manner. Several Nigerian missionaries felt called to work in dispensaries around the lake shore, and they would treat a wide variety of ailments, sorting out surgical and other serious cases for the doctor to see on his lake visits.

Alongside all this medical work the Christian Gospel was explained and taught to the people by Dr Carling and his fellow missionaries, both African and ex-patriate.

In its early beginnings their efforts were limited by difficulties of transport both on and around the lake, and by the lack of dedicated personnel. As time went on it became evident that this work was necessary and effective at both medical and spiritual levels. Additional African workers joined the team as missionaries, including some Tchadians. A launch was provided by Christians in Britain: this was Albishir. More dispensaries were opened at points all around the lake and on some islands, and small groups of converts formed themselves into simple churches. By 1967, however, even Albishir was proving inadequate to the needs of this growing outreach.

Earlier the Mission Aviation Fellowship had joined in this work by providing an amphibian plane *(See: Plate 32.1),* the one which had rescued us when the *Jolly Hippo* broke down, and this had greatly extended the doctor's scope. Now he envisaged a larger hospital boat, with a very shallow draught - not more than about 50cms - a triform hull, engines which could be raised from the water for inspection, and, above all, an operating theatre.

At the time of writing (1971) that vessel *Albarka,* built and purchased with the gifts of Christians from all over the world, is on its way to Lake Chad.

When it goes into service, it will tour the lake on a regular circuit, while the doctor will fly from dispensary to dispensary, examining patients and then operating in the floating hospital. Post-operative care will be in the hands of the African missionary dispensers and nurses so that the patients can remain in their own home areas and near their relatives during convalescence.

Such is the work of this remarkable man and his dedicated team of helpers. He has clearly demonstrated to them his skill in handling boats, and his willingness to exhaust himself for their sakes. His reliance on his African missionary fellow-workers is also a powerful influence in this work.

It is in the efforts of this small, but effective team that I see, epitomised, the values and motivation that are the only genuine key to safe development, appropriate to modernisation, and real enlightenment, on and around Lake Chad.

We may ponder the factors that have moulded Lake Chad through its ancient history, with all its unique characteristics and hydrologic peculiarities, up to the present time, attempting to view them in their broadest perspective. We may consider and try to assess the intensity with which modernising commercial, technological and political pressures are now being exerted on both the environment and the people.

We may admire the attempts already made to ensure that these pressures are regulated through cooperation, coordination and communications at all levels.

*Nevertheless, it seems to me that the status quo of the whole Lake Chad complex is too delicately poised for complacency. On the one hand lies its encouraging potential for prosperity and progress; on the other its alarming potential for disaster. Whatever the result, it is quite clear that, today, man feels he can and should try his hand at harnessing and mastering Lake Chad. I wonder which will persist in the future: lake.....or mirage?*

*Chapter 10 - Modernisation and enlightenment*

# PART IV

# APPENDICES 1 and 2.

# THIRTY YEARS ON

# 1973 - 2002

# Chapter 11

# 1974: The Fragile Lake

On 15th October 1973 I received a letter from a visitor to Lake Chad who wrote *inter alia* 'On a recent tour in Northern Nigeria I took your book with me to Lake Chad. When I arrived at Baga where the lake is supposed to be, I found that it had disappeared, and the new Government Rest House, purposely built close to the lake for intending tourists, forlornly situated in a sandy waste. Fortunately I was able to borrow a Landrover to follow a deeply rutted sandy track that finally led to the lake shore....A fisheries officer accompanied me and told me how concerned the fishermen were becoming as fish were so small and there was actually a lack of fish in some places. The fears expressed in your book about over-fishing are therefore becoming more acute'.[44]

In March and April 1974, a companion and I took a 4.5m launch, the *Merry Mermaid* (on loan from the Jos Wildlife Park Manatee Project) powered by a 20 h.p. outboard engine, to Lake Chad. How had the lake been affected by the acute drought of the past two years? And, in turn, how had the life of the lake people been affected?

**Exploratory flight over the whole lake area.**

Friends were pessimistic about the likelihood of our being able to cruise at all on the lake even with a launch drawing only 25 cm draught, so we decided to fly over the lake first to assess the situation as best we could. In March and April, very dry months when the heat is on the increase, a strong north-easterly wind called the *Harmattan* carries fine diatomaceous desert dust which fills the atmosphere over most of the Chad basin and reduces visibility severely. This interrupted our first flight but we were able to complete it next day, and even to land on the sandy, bumpy runway of one dune island in spite of a stiff, dust-laden breeze

The flight gave us an important insight into the condition of the lake. Although there was quite a large area of open water in the south basin just off the Chari River delta, the remainder of the south basin appeared to be completely covered with a mat of floating vegetation. Through this mat we could see that the fishermen had cut channels just wide enough to take their largest canoes *(See Plate: 25. 1,2,4)*.

A large circular pool and another crescentic clear area seemed to lie midway between Baga and the open waters and seemed to be connected by man-made channels. Baga itself lay some 5 km. away from the water's edge, and the fisheries and the customs and excise stations at

## Chapter 11 - 1974: The Fragile Lake

Portofino, our mooring base on my 1969-70 expedition, seemed to lie, high and dry, very far inland.

We turned and flew northeast and followed the 'Great Barrier', the wide sand bar that marks the division between the north and south basins. This bar was in 1969/70 covered with open lagoons and floating papyrus and *Phragmites* rafts or 'islands', and navigation across it was not difficult. But now it was partly dry and partly covered with shallow water and short grasses. A motor track wound across it linking Baga and Baga Sola, but at that particular time it was not in use due to several wide, muddy patches that made it impassable.

During this visit the significance of the Great Barrier, as it was directly affecting life of the two basins, had become apparent. With decreasing overall rainfall on the whole Chad catchment over the past ten years, the average annual lake level had been steadily falling. In 1973 when the water level fell below the level of the great Barrier, and the Yobe River was no longer delivering any water at all into the north basin, the level of the north basin dropped even lower than that of the south basin where the Chari River was still delivering sufficient water to balance losses by evaporation.

During the 1973 rainy season, the Chari and Yobe rivers again delivered sufficient water to raise the level of the whole lake just above the Great Barrier, but in 1974 not only had the Yobe again failed completely, but even the

Chari was running at a record low. The result was that the south basin was maintaining a constant level equal to the top of the Great Barrier, but the level of the north basin had, by April 7th 1974, fallen 1.50m lower than that of the south basin, and was still falling rapidly as a result of an exceptionally high evaporation rate. It is also interesting and important, in view of its practical repercussions, that the bar surrounding Bol proved to be higher than that of the Great Barrier, and the part of the lake around Bol was cut off as a stagnant pool even before the north and south basins separated. Only three years before, irrigated polders around Bol were richly producing wheat, rice, sorghums and vegetables. Now they were just a barren waste.

At Bol we visited the Station of the Swiss Branch of the Sudan United Mission. The missionary met us at the airstrip amid a swirl of sand and dust, and swept us away for a meal and a visit to Bol town. What a scene of desolation: just a few rows of sandy-grey, moorish-style buildings set in soft thick, dry sand! There seemed to be no shade anywhere: only the white glare of the sun filtering hotly through the gathering harmattan haze. We asked if we could see the original depth guage set by General Tilho to record the changes of lake level. He recorded the lowest-ever minimum in July 1907 when the level fell to 279.87m above sea level, the Great Barrier was passable on foot, and the north basin was completely dry north of the latitude of Bosso.

## Chapter 11 - 1974: The Fragile Lake

Unhappily we learned that the historic guage had been removed. A number of derelict steel barges and launches lay stranded, deeply embedded at different levels, on the steeply sloping strandline, a strandline reflecting overall changes in level of almost 5 meters during the present century. The waters lapping the shore looked murky and there was no passage out through the low rim of vegetation surrounding the ancient harbour. As we took off again over Bol we could see a series of stagnant pools between it and the open waters of the south basin. These had gradually dwindled, we were told, and had first become covered with green algae. The cattle could then still drink from them. Then came a red alga, poisonous to the cattle, which caused heavy losses until the herdsmen took to digging shallow wells for them and watering them from gourds.

Our missionary friend described the effects of the drought on the people. The bulk of the indigenous pastoralists had suffered heavy losses of livestock and most had trekked away south into the southern parts of the Tchad Republic, Cameroun and Nigeria. Inevitably some of their old folks had remained behind. The trading and farming communities had become divided and some had emigrated southwards. Hardest hit were the very old, the very young, and the sick. It was to these that the missionaries had been able to distribute relief supplies on a selective basis. And it was some of these who had responded not only to the physical care shown them but also the message of God's eternal love through the Gospel of Jesus Christ. These cheerful missionaries certainly led no easy life at Bol, yet we had the impression of people who were happy and satisfied in the work to which they had been called.

**Launching our workboat 'Merry Mermaid' at Baga.**

Having completed our air reconnaissance, we now launched the *Merry Mermaid* at the lake shore outpost of the port of Baga, (named 'Dorowan Gowon' after a visit by the head of State of Nigeria), a large rambling temporary village built entirely of grass mats tied over wooden poles. Scattered, short papyrus plants among some unhappy looking Vossia grasses marked the lake shore, while stubby ambach stems poked up through the shallow water off-shore. How different from the 1969/70 expedition when Baga stood right on the shore-line!

A large fleet of dugout and plank-built canoes of various sizes lay stranded on the muddy ooze, but there was not a single reed *kadai*. Large, heavily laden freight canoes came in at intervals during the evening, forced over the shallow mud by floundering boatmen and passengers. Each canoe was loaded to capacity with bulky cartons of dried catfish, and immediately it reached the shore, a group of businessmen would gather and begin bargaining vigorously, while the sellers hastily off-loaded the cartons on to the shore. Soon a landrover or a lorry would back down to the shore, and then, the bargaining completed,

## Chapter 11 - 1974: The Fragile Lake

would load up ready to carry the fish to Baga town, and thence onwards by huge articulated lorries to towns all over Nigeria including Kano, Ibadan and Enugu *(Plates 22.2;24.4 and 5)*.

We prepared our launch carefully, not knowing quite what to expect or how long we should be afloat, and late next morning we were pushed out over the mud into the navigation channel. Then we were pushed and poled almost 5 km before we could start the outboard motor. We had engaged two rather modern-looking Yedina boatmen, father and son, who both enjoyed the name 'Adam'. Once we got the engine going they climbed aboard and briefly relaxed.

Unfortunately the circular area of open water that we first encountered was only about 3 km across, so we were soon struggling again in a narrow navigation channel. Then we broke into the long crescentic stretch of open water that we had seen from the air, leading towards the remnant of open water of the south basin around the Chari delta. Finally this gave way to a series of very shallow, narrow lagoons. So far, navigation channels had been cut through the floating mat of mauve convolvulus *(Ipomoea rubens)* plants, through which scattered ambach saplings poked to a height of between one and two metres above water. It was very sultry within the navigation channels and we saw very few birds.

*Lake Chad versus The Sahara Desert*

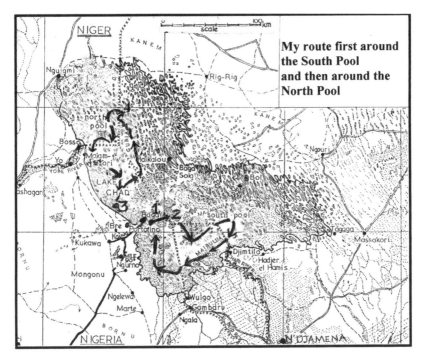

**EXPLANATION:** *NOTE. The route of the "Merry Mermaid" launch, and the details of floating vegetation, sand (dune) islands and lake shore, are only approximate due to the unstable surface levels of the lake.*

1. Dorowan Gowon (temporary "port" about 5 kms from Baga, passable only by 4-wheel drive vehicles by way of sandy tracks,).
2. Kangalom dune (or "sand") island straddling the international boundary between Tchad Republic and Nigeria.
3. Kangaruwa. A temporary "port", like Dorowan Gowon, built entirely of grass or plastic sheeting huts, set on the former lake bed.
4. Kinjeria, formerly a lake-level metering station on a dune island, now in 1974 with the meter standing high and dry.

*14: My route around the lake in 1974.*

Nearer to the open waters, however, there was less ambach, and here we saw large numbers of white herons *(See Plate 7:5),* night herons, west african terns, and a few pied kingfishers. After leaving the open waters and entering more lagoons, we encountered rosy pelicans in large numbers. These lagoons were flanked with stands of *Vossia, Phragmites* reed-grass, a few short papyrus stands, scattered reedmace and some very fine thick stands of tall ambach *(Plate 7.6).*

Soon after 5.0p.m. our engine failed and needed some detailed attention. We struggled with it for a while and then decided that it would be essential to make some sort of camp before the mosquito hordes arose at dusk. It was quite clear that there were no floating camps anywhere near, and Adam Senior assured us that it was still a very long way to *terra firma.*

Moreover, he also explained that it was now quite impossible to make a floating camp as the condition of the *Phragmites* and papyrus simply could not support our weight. How different from my previous visits when all we had to do was to step out of the boat carefully on to a floating island, slash down a few Phragmites or papyrus fronds to form a stable mat, and then cook our food and lay our mattresses on it. A few fronds tied overhead would support the essential mosquito net.

In spite of my companion's misgivings, I insisted that we make our way forwards, the two Adams pushing while my companion and I poled to a tall, thick ambach stand. Here Adam Senior felled some poles across our bows to form a platform on which we could cook supper. I then set up my mosquito net and laid my mattress on the damp, insecure platform, while the three men bedded down in the cock-pit of the boat. We were only just in time before myriads of zizzing mosquitoes enveloped our mosquito nets, maintaining their humming like a swarm of bees until the sun rose again.

Next morning, having repaired our outboard motor, some six hours of cruising, interspersed with periods of poling, brought us via open waters to the sand island of Kangalom. This island is actually within the Tchad Republic, but it had become inhabited near the shore by Nigerian Kanuri people, and further inland by Yedinas. The latter said Kangalom was in Tchad Republic, while the former declared it to be in Nigeria. We encountered the same kind of difference of opinion as to where Tchadian and Nigerian territorial waters began and ended in the north basin also.

As Kangalom is some 30 km from Baga, and as it was impossible to find or make any intermediate floating camps, the local people would begin loading their freight canoes at Kangalom very early in the morning when the sun rose, so that they could reach Baga with steady poling before nightfall. Here, as everywhere else on the lake this year we were repeatedly struck with the busy activity of the

## Chapter 11 - 1974: The Fragile Lake

lake people. In the south basin catfish were extremely abundant and evidently easy to catch in gillnets and on foul hook lines, so, while the freight canoes prepared to leave for Baga, a fleet of smaller dugouts, plank canoes and 'swimmers' on huge gourds set out for the fishing grounds *(See Plate 12: 2,4,5)*. Some of the canoes travelled as much as 15 km to where their nets had been laid the previous day.

This was a completely different way of life from that seen on my previous trips to the lake when the fishing was all done in close proximity to the floating camps, and when plank canoes and gourds had not yet replaced kadais, and were, in fact, only rarely seen near the shores at Baga and Portofino. This time we encountered only one *kadai* during the whole trip, but we saw gourd boats everywhere in the south basin.

### At Baga again.

Our homeward journey back to Baga was aided by a stiff following breeze, and especially on the open waters we made fair speed. Many of the canoes had a crudely rigged mast with a cross bar and a blanket loosely tied to catch the breeze, but I never encountered any kind of yacht rig for more sophisticated wind-use on Lake Chad other than that of my own sloop, the *Jolly Hippo* used on my 1969/70 expedition *(See Plate 12: 3)*.

We had decided to use the government 'Catering Rest House' at Baga as our base during this visit to Lake Chad

Many features of this institution aroused our curiosity and comment. It consisted of three duplex breeze-block chalets, two prefabricated wooden chalets and the main office and catering block. The whole compound lay two meters below the well-known lake level maximum and it was interesting to speculate as to its position during a future 'Great' or 'Medium' phase of the lake's level.

Moreover, under a huge tarpaulin, lay a brand new 'Everglades' 10-seater air boat. We were informed that 'this has been imported for V.I.P. tourism'. We wondered about the probability of such tourism actually ever becoming a reality and about this boat's suitability, for it could not possibly ride the water at the lake's present 'Little Chad' phase, and during the 'medium' and 'great' phases it would be unstable on the choppy and rough open waters.

What, we asked ourselves, would be the ideal boat for future Lake Chad tourism? Would motor launches or speed boats be suitable? They are very noisy and as long as gillnet and foul hook line fishing continued, would be constantly getting entangled. Would dugout canoes be suitable? They are very uncomfortable and unstable. Yachts? Certainly in the 'medium' and 'great' phases of the lake, yachts with a reasonably shallow draft, fitted with fish-net deflectors, would be delightful.

But, assuming that the lake **may** rise again some day, why not the *kadai?* It is traditional, unsinkable, stable, and can be made in any size. It is silent, picturesque and romantic, and what could be more enjoyable for the tourist than bird watching among the lagoons, or poling across the silken lake waters, in a *kadai?* It might be practical to suggest that waterproof over-trousers be provided for the tourists as 'seating' inside a *kadai* is actually in water!

## Transfer of 'Merry Mermaid' from Baga (south basin) to Kangaruwa (north basin).

We now negotiated with a wild lorry crew to transfer the Merry Mermaid from Dorowan Gowon in the south basin to Kangaruwa, another temporary town, on the shores of the north basin. It was necessary to obtain a 4-wheel-drive lorry to negotiate the 49km of deep sand. The drivers of such lorries and Landrovers have exploited the upsurge of fishing during the current reduction of Lake Chad and carry on a most lucrative business at inflationary rates. Nonetheless, we finally managed to agree on a price and the boat was safely conveyed to Kangaruwa and deposited on the muddy lake shore.

Like the local people we quickly had a grass shelter built to house our camp guards and ourselves, with the car alongside, some way above the 'damp line' on the shore. Here we dug a shallow well from which to draw drinking water. Although this was only 60 cms deep when dug, the lake level dropped so much during our week's stay that by the time we left, we had to dip a whole metre's depth to reach the water.

Kangaruwa was a huge village built, like Dorowan Gowon, entirely of grass mats tied on poles. There were no sanitary arrangements and we were not really surprised to learn that it was currently the focal point of a raging cholera epidemic. The Health officer in charge of the vaccination and treatment centre was unable to give us figures for the number of deaths as he said that, owing to the great distances and slowness of transport, the majority of patients died either on the way, or in their villages on the islands. We did see one such patient being brought in to Kangaruwa in a canoe in which a rough shelter from the sun had been constructed.

Leaving Kangaruwa, our launch again had to be pushed for about 5 km over sticky, very shallow mud. This was really hard work, and several times we had to enlist aid from the crews of other stranded craft. Once we could use the engine, we made good progress, keeping an ever watchful eye open for fish nets and lines which would entangle the propeller. In some places the concentration of this tackle was quite incredible and it was very difficult to avoid it. Unfortunately the nets and lines were usually only marked by two sticks, one at each end, so, where they were particularly numerous, it was hard to identify their positions. There were also stretches where the water was so shallow over the mud that the engine laboured badly and we could only make slow going. We set our course north-north-east in hopes of reaching an island called

Kinjeria in the Tchad Republic. This we knew was in the true Yedina area and we hoped to find typical villages and cattle.

On our way we spent a night on a very barren, low dune island. A deserted village of clay huts lay on one side of the island. These huts were in the Hausa style. On the other side was an untidy encampment of grass huts, and this we discovered was a Jukun camp. The Jukun are fishermen native to the Benue River and this was my first realisation of the character of the invasion of Lake Chad by 'foreign' tribes which had taken place in the previous two or three years. The Jukun fish with the usual fixed gillnets and lines, but in addition with a very long net called *taru*, sometimes up to 200 m long, which they draw through the waters of selected pools enclosing vast numbers of fish. These people were friendly and hospitable and were ready and willing to demonstrate their fishing techniques to us.

**Lake tribes.**

Continuing northwards we passed numerous low, flat, utterly barren sand islands. A few of these were narrowly fringed with low, green pasture grasses where herds of cattle grazed. On some we saw several camps and villages, but it was apparent that these represented mixed tribes: Hausa people from Sokoto and Kano, Jukun and Nupe tribesmen from the Benue and Niger rivers, Kanuri and Kanembu fishermen from the surrounding mainland,

and, in the minority, the true natives of Lake Chad, the Yedina. Each tribal group had its own encampment or village clearly separated from any neighbouring ones. It was hard for me to believe that the formerly fearsome Yedina, known to the French colonists as 'the pirates of the papyrus' could have so completely yielded the fishing, pastoral and living rights on their magnificent lake to such a conglomeration of 'foreign' tribes.

Early this century, sorghum was introduced to the islands and a few Yedina families attempted some cultivation of this alongside their traditional pumpkins. With growing commercial activity around the lake, the Yedina began to trade in manufactured goods and adopted some of the more convenient Muslim customs and beliefs.

During the past fifteen years medical missionaries set up a number of outstations with dispensaries and small churches, bringing the Christian Gospel to some of the island people.

Formerly the Yedina were the true masters of Lake Chad, and few mainland neighbours ventured far among the islands. On my various visits over 19 years, including this trip, I always found the Yedina to be courteous, friendly and hospitable. On arrival at a Yedina village, the headman and elders would walk down to the shore, receiving their visitors from the boat with a smiling welcome and the presentation of gifts of fish and milk.

## Chapter 11 - 1974: The Fragile Lake

This was still true on our present trip, but I did not have the same sense, as in the past, of a closely integrated, secure village community.

Indeed, I found it incredible that in only four years their way of life had lost so much of its distinctive traditional character. True, they still lived in typical portable grass (but fewer papyrus) huts, but now these had been repaired with sheets of blue plastic material and with mats imported from the mainland. True, they still milked and pastured their cows as before, but now the milk was stored in plastic buckets and gaudy enamel bowls instead of in finely woven fibre baskets. The women pounded imported guinea corn with imported pestles and mortars, and cooked tea in imported aluminum kettles on imported iron braziers.

We camped close to Yedina villages and noticed their activities as they went about their daily tasks, but all the time I sensed that they had been swallowed up by the invasion of foreign goods and foreign tribes. Even the herds of cattle were no longer the pure *kourri* strain, but were adulterated by cross-breeding with the *Bornu Red (See Plate 21.6)* and *Fulani White* varieties. It seemed that, like the Yedina themselves, even the cattle were fast losing their identity.

All along the way we asked different people what they thought about the desiccation of Lake Chad. The recent immigrants did not expect it to rise again and were just cashing in on a golden commercial opportunity while the fish were heavily concentrated in the diminishing pools. They did not care whether they would leave enough fish for future requirements. But the older Yedina men said they 'knew' the lake must certainly rise again as it had always done in the past, only they hoped it would begin to do so soon as their greatest present hardship was the shortage of grass and papyrus on which their whole life depended.

They complained that the 'strangers' were taking too much fish and that there were no big fish any more, but, none the less, they still had sufficient both for their own needs and for sale. Any delay in the recovery of the lake would also mean an acute shortage of pasture for the cattle, now only just sufficient to sustain them; and due to the reduction in the formerly abundant papyrus and *Phragmites* stands they were having acute difficulty maintaining their houses adequately to withstand the possible high winds and sharp storms that could occur during a future 'good' rainy season.

We reached Kinjeria island on a Sunday morning. The mission pastor had held his morning service before our arrival, and he, with the village elders, came to welcome us ashore as we landed. He held another service in his simple grass church which we attended that evening. In the afternoon, when the burning midday heat had passed a little, we strolled up to the village.

A hut was to be transferred nearer to the shore in the wake of the receding lake. About thirty men, women and teenagers had gathered. They brought some strong forked branches and thrust these into the hut walls. There was the usual initial discussion and some jokes were cracked. There was laughter. Then, with a rhythmic song, everyone bent to the work, and the hut was lifted and carried down the narrow way between two other houses. It was in the prevailing state of disrepair and was very lop-sided, but happily, on arrival at its new site, it was pushed into shape, somewhat restored, and everyone seemed pleased with it *(See Plate 17.4).*

**Return to Kangaruwa**

Our return journey to Kangaruwa followed a more westerly course and we frequently encountered shallows and sand bars off the Yobe delta. Fishing in this area was even more intense than anywhere else we had been and the lake was silky calm in the intense heat. In the midday glare it became impossible to distinguish the water from the sky and piloting the boat was difficult as I felt I was the victim of a hallucination in a world with no horizon *(See Plate 24.3).*

Approaching the shallows off Kangaruwa we saw the only hippo of our whole trip. He was resting on the mud in a shallow place. Unfortunately we were unable to approach closely due to the shallowness of the water and the sticky muddy bottom. We had been told that wildlife officers of the Forestry Division in Maiduguri had exterminated all the hippos in the western part of the south basin and that only isolated individuals and small groups now survived in the north basin. A few were said still to exist in the lower reaches of the Chari River. The elephants were said to be doing well in the swamps south of Baga but their habitat along the lake shores was fast being destroyed by cattle. We saw no otters or sitatungas on this trip.

Since we set out a month earlier, the lake level in the north basin had fallen some 25 cm, and this time we had to have the launch manhandled over the mud for about 8 km. We were exhausted at the end of this trip. The intense heat (up to 125°F) had been very trying and there had been no escape from the sun. Everything spoke to us of harshness, of the struggle for survival, of lost identity, of insecurity, and of greed.

*I felt that an era had passed and even when the overall lake level rises again, as hopefully it should in the next year or two , it seems unlikely that the Yedina will ever fully resume their former way of life as the 'people of the grass', the 'pirates of papyrus'.*

---

\* See ch. 12: it has not risen again significantly up to now (2002)

# Chapter 12

# 1973 to 2002

## 1. Astronauts notice that Lake Chad is seriously dwindling.

One day in the autumn of 2001, I was half listening to the radio and reading a book at the same time when I heard the words 'Lake Chad'. Immediately alert, I found I was hearing the voice of an astronaut up there in space saying that he was then orbiting over Lake Chad in the Sahara desert but that it looked very small these days. He remarked that in the past it was a huge expanse of water and an easy landmark seen from space. I was immediately galvanised into action. How small? Still receding? Recoverable? What about the people, the fishing, the *kourri* cattle, the papyrus and ambach, and the wildlife?

I was soon in touch with the NASA Goddard Space Flight Center from which I was sent the amazing 'Landsat' images on *Plate 26,* covering the period 1963 to 2002. Research into recent literature from the IRD\* (formerly ORSTOM), in Paris and the Lake Chad Basin Commission in N'Djamena confirmed that the lake had never recovered since the awful drought of 1973/74. In fact, by 1995, it had

---

\* Institute de Recherche pour le Développement

steadily receded to surface levels far below any previously recorded in the 19th and 20th centuries.

I had never published the account of my visit to the lake in 1974, when it had already begun its dramatic decline, and when I personally saw and experienced those early changes. So I felt that an account of the 6 weeks spent then on the lake should be published now *(See Chapter 11)* describing my experience of the early days of the lake's decline).

*Diagram 6 (Page 119)* was the graph I constructed from data made available then by ORSTOM covering the period 1870 to 1970, and acknowledged in my original book 'Lake Chad' (1972).[45] Where the ORSTOM records were incomplete, they were supplemented by data of the rise and fall of the Chari and Nile Rivers, which followed a similar pattern. A comparable pattern has also been seen to some extent on Lake Naivasha in Kenya.

The graph shown here in *Diagram 22* is constructed from a copy of my former graph[46], now extended on the basis of more recent data supplied by the Lake Chad Basin Commission. It does not exactly coincide with that published by *Olivry et al*[47], but does demonstrate the trend in the lake's surface oscillations. When Jane Sutton re-visited the lake this year she measured the lake surface level by GPS* above mean sea level and recorded the level of 280.7m on 19th January 2002. This was at a time

---

* Global Positioning Satellite

*Chapter 12 - 1974 to 2002*

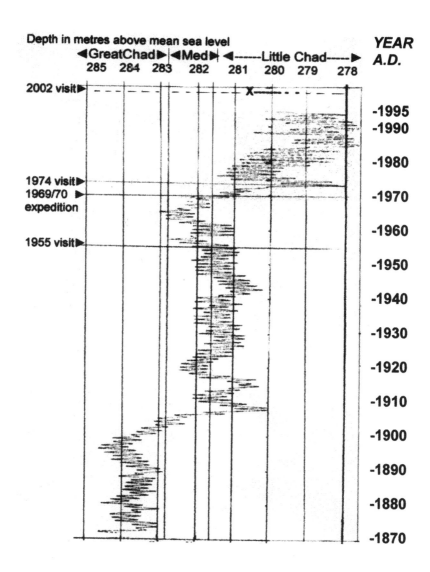

*Diagram 22. Average oscillations above mean sea level of the surface of Lake Chad 1870 to 1995 drawn from available data courtesy LCBC., with an actual measurement **(X)** on 19 January 2002 at 280.7 m.(Sutton)*[48]

when the lake *should* have risen to its maximum for the year, but the reading cannot be compared with the earlier part of the graph which shows *annual averages*.

She was then shown on the current Bol meter that the lake had, by that date, already begun to recede and had passed its maximim by a month, so any hopes of a further rise in early 2002 were already dashed. This observation was confirmed by a resident pastor who stated that the water had again come within 5km of Baga township.*(See ref. Chilvers et al, page 348).*

I was accused by the editor of a well-known geographical magazine in 1974 of being *'obsessed'* with the *'oscillations* of *the surface level of Lake Chad'* rather than with its then *'living'* situation! However, I believe that informed global opinion in 2002 is all too aware of the facts of climate change affecting very seriously and practically the availability and accessibility of water to millions of people world wide. Lake Chad is a very obvious example of the effects of that change, not only visibly, but also measurably. It is thought that about 8.3 million people are today affected directly by Lake Chad's decline, and many more indirectly by the reduction of its wider produce distribution.

Headlines 'scream' about the importance of the falling level of Lake Chad: 'Lake Chad disappearing'; ' Africa's Lake Chad shrinks by 20 times due to irrigation demands and climate change'; 'A shadow of a lake'; 'African nations seek global help for Lake Chad'

## 2. Can Lake Chad be artificially replenished?

So important is that change that the Lake Chad Basin Commission is seriously considering a scheme to re-charge the lake artificially by building a canal from the Benue River above the Gauthiot Falls in the Central African Republic to join the Chari River south of N'Djamena. This is because it has been feared for many years that there was a leakage across the Chari/Benue watershed into the Benue River robbing Lake Chad of a significant percentage of its potential re-charge. It is said now to deliver only 50% of its 1930-60s inflow. How much of this leakage is due to human activity on that watershed is not clear. The scheme is as yet only conceptual[49].

There is no question that on the opposite side (northwest) of Lake Chad where the Yobe (Hadejia) River used to supply a significant inflow to the northern basin, the building of a huge dam (Taiga Dam) in its upper reaches near Kano and extensive irrigation schemes along its course, together with climate change, have actually *nullified* its flow into the lake. The use of irrigation (by means of bore holes and wells began on a large scale in 1963 and diverted flood water and unconfined ground water) quadrupled compared with the previous 25 years. This has undoubtedly contributed to the dramatic effects of climate changes occurring since 1973, which have resulted in the severe regression of the lake from an area of approximately 25,000km$^2$ in the 1950s and 60s to an average of about 2,500km$^2$ in the period 1973 to1995.

It is true that seepage pans after the rainy seasons do still form, and in some places the water table can still be tapped by wells, but the desert has virtually extinguished the northern basin. Pastoralists did, for a time, burn off extensive areas of vegetation around the desiccated northern parts of the old lake bed, hoping to bring on fresh growth, but vegetation for livestock around and across the northern basin became so scarce that many of the people have moved away. The cattle remaining on and near the lake are more obviously of mixed breeds and fewer in number than previously, mainly owing to the reduction in available fodder. *Kourri* cattle, as a pure breed, have become quite rare on the islands and mainland shores compared with their former dominant position there *(Plate 30.3)*.

### 3. Water management.

Irrigation projects around Lake Chad have long been a contentious issue, especially when the lake fails to remain at a relatively stable level. Several failed primarily due to civil strife resulting in inadequate management. The most obvious example, however, of management ineptitude, is the 'Southern Chad Irrigation Project' (SCIP) which had a goal of irrigating 67,000 hectares (about 260 sq miles) with an average cropping intensity of 130%, and resettling about 55,000 farming families on to the irrigated land.

This was in addition to those Nigerians who had already adapted to the lake's retreat by moving on to lands within the former lake bed, so that villages within the Nigerian portion had already risen from 40 to about 100 between 1975 and 1979.

Although a pilot project had been run in 1962/3 when the lake waters were high, it was not until 1979 that the major project was commissioned, with the installation of pumps and canals to carry water from the lake shore to the farmers' fields. It only lasted 6 years with only 7000ha irrigated. Water seeped out of the unlined canals, pumps failed, and when water did reach farmers' fields at all, they over-reacted by siphoning and/or breaching the canals, water-logging their fields and losing large portions of the growing crops. It simply did not work productively at the lower lake levels.

The on-going use of irrigated polders for the development of adapted varieties of cereals (notably wheat, maize, rice, and vegetables) was started many years ago around Bol (in the northern archipelago of the southern basin in Tchad). On the Nigerian side around Baga (on the south western side of the southern basin in Nigeria), and Malamfatori direct irrigation from boreholes and pumps was stepped up.

The polders around Bol have been further developed with control dykes or barriers by SODELAC (Société du Développement du Lac) with 3000ha under cultivation since they began in 1963, and more recently by SARTOM (a French company) on a further 8 polder barriers *(See Plate:28.2).*

The produce was formerly shipped to N'Djamena by barges, but since the extreme decline of the lake level, and the civil strife, export has been by lorries - an expensive and difficult method. SARTOM have plans to dredge a shipping channel from the polders to the Chari river mouth. However the value of produce from the polders is said to have diminished from around 800,000 tonnes to a mere 370,000 tonnes, so perhaps the value of opening of a channel may be in doubt. Tenant farmers who have plots in the polders are undoubtedly at an advantage by having access to the controlled water supplies.

*Diagram 23. Maize (Logosrum variety) and wheat (Duolo variety) grown on SODELAC polders. near Bol: The wheat is bearded like barley and protects the ears against birds.(Print 2002, Jane Sutton)*

Another major problem, particularly with irrigation in the northern archipelagos, is the intensive deposition and downward seepage of concentrated natron whenever the lake rises at all, especially during its lower level cycles over its inundation areas and island shores. This seriously damages the soil, which then requires treatment before cultivation. Not all this natron is commercially usable or even accessible.

Where old-fashioned irrigation is possible by means of *shadufs,* farmers can still grow useful small-holding crops for home and cash sales. Pumps powered by solar panels are realistic for individual use and for small communities, but are not yet widely in use. Unfortunately even the solar powered satellite instrument for measuring water levels at Bol was vandalised earlier this year (2002), so there may be resistance to investment in this technology for general use for some time.

Within the polders near Bol farmers can each rent half hectare plots, in return paying a levy of 30% of their wheat crop for the use of tractor and pump. They build up 'walled' squares using hand hoes and flood them twice weekly. In these polder farms they grow Italian wheat, sorghum, tomatoes, lettuce, potatoes, carrots, courgettes; and, elsewhere in some polders and gardens, also fruit trees (mangoes, guavas, etc).

A very interesting study was made by Dr Terri Sarch and Charon Birkett[50], into the fishing and farming situation of settlers on the south-west lake shore of the southern basin, in the vicinity of Baga. It included Kanuri and Hausa, and a few Fulani, Shuwa and Yedina households and was based on a survey conducted in 1993 and further research in four villages on the lake shore in 1995. The study concentrated on a livelihood analysis in the wider context of environmental change and development, relating it specifically to intra-annual lake fluctuations rather than the long-term lake level cycles.

Interest focussed on the seasonal fishing and farming activities of the households listed above, and on the flexibility of their response in utilising available resources on their own initiative, which included their willingness and ability to move house towards or away from the lake shore according to the lake's seasonal rise and fall.

Rise in lake level translates into floods which are exploited by a range of fishing methods adapted to different flood stages. The subsequent falls in lake level are then exploited by farmers taking advantage of both residual soil moisture and fertility and the prevailing hot season. Sarch and Birkett say: *'the fishers and farmers of the lake shore are doing better than coping'* and give data of the economic returns on the saleable produce and the very adequate self-sufficiency levels as regards household requirements.

In trying to obtain reactions from people now living within the dried-up lake bed and near its former shorelines it has become apparent that answers to my own questions have mainly come from a *new generation* of lake dwellers. These *only* know the lake in its present 'Little Chad' state, and seem naturally to have a simple acceptance of the limitations on lake travel by boat through today's lengthy hand-cut channels and hand-dug canals, the new vegetation patterns, new food habits and the more modern types of boats, houses and equipment.

Many are very conscious also of the instability that came as a result of the civil wars and brutality of rebels, but do not appear to regard the lake's current condition as life threatening. It seems that even the Yedina of the northern archipelagos and the open waters of the southern basin still remain mainly distinct and thriving, but in the southern areas have acceded to the dominance of the Kanuri tribe, inter-married and use the Kanuri language readily. Chilvers *et al.*[51] Their needs seem to be met, in the main, without dependence on papyrus for houses and canoes, or palm fibres for woven pots, or a special local breed of cattle: anything will do!

The markets and towns have grown and imported manufactured mainland goods, but of course the cost of living for all lake and shore dwellers *per capita* has risen. As Sarch and Birkett point out, their exploitation of the alternating lake bed fertility for wet season farming and dry season flood fishing is adequately balanced against their family needs for the time being.

## 4. Towns and villages.

*1. Education.*

The villages and former small towns have in fact grown in size, notably, for example, Baga, Bol, Baga Sola, and Djimtilo. On the Nigerian side there are village and town primary schools with secondary education, trade school, and university in Maiduguri.

Education at primary level in the Bol *préfecture* is available on some islands, on which there are some 800 villages with about 25 Yedina schools. At Bol, the grand new secondary school or *Lycée* is shown compared to the former old building in *Plate 29 .1 and 2.*

Jane Sutton's enquiries[52] on today's education in the northern part of the lake around Bol revealed that the Lycée, completed in 2000, is the only two-storey building anywhere around or in Bol, and in January 2002 there were 692 boys (of whom about 25% were of the Yedina tribe, the remaining 75% comprising mainly Kanembu together with other tribes), and 160 girls. Teachers numbered 22, with 5 assistants. Sport includes football. This education requires 7 years to obtain the *baccalauréat.* Students seeking higher education can proceed to university in N'Djamena or to the college in Niamey, Niger.

An additional school outside Bol caters for the schooling of specifically Muslim girls. The Bol primary school started in 1959 and this year (2002) has 757 boys and 355 girls in 12 classes, with 16 staff comprising some Muslim teachers and others from southern Tchad.

Sutton also reported that the congregation attending the Christian (evangelical) church in Bol, comprised mainly people from places in southern Tchad. She noted that workers from the 'Summer School of Linguistics' (SIL) were working as a non-government organisation (NGO) in association with the Bol *préfecture* on a literacy project for the Yedina tribes-people.

*2. Around Bol town and offshore.*

She was able to spend some time in and around Bol township meeting people and asking questions. She saw the thriving market, the camel park, the waterfront, the veterinary department, forestry department (tree plantations are important as wood is still used for cooking and shade trees play an important role in the town itself), and small holdings near the lake shore. She stayed at the TEAM* compound in Bol, and from there, was able to hire a canoe for travel out to several islands in the archipelago, see something of the fishing, meet various islanders and, on one island, the local *Imam* (Muslim priest) at the island mosque *(See Plates 30-31).*

---

* The Evangelical Alliance Mission

Even travelling from N'Djamena to Bol by road at the beginning of her visit, was important because she saw how desperately dry the sahel (in places now genuine desert) had become, with high bare sand dunes as they approached Bol. She noted that the camels of a passing camel train, transporting goods, probably from farther north, were very thin.

On her return flight to N'Djamena at the end of her visit, the MAF pilot enabled her to take stunning pictures of the only rocky outcrop in the vicinity of Lake Chad, the Hadjir el Hamis *(See Plate 27.2).* This, when enlarged, also shows, at the lower right corner of the picture, a village in a very barren sahelian landscape near the foot of one of the rocks.

5. Medicine.

On the Nigerian side there are dispensaries in some mainland towns, villages and islands, a well established mission hospital (Action Partners) in Molai, and a government teaching hospital in Maiduguri.

N'Djamena has a large government hospital, and there are clinics in certain towns neighbouring Lake Chad. Bol has a government hospital presently (2002) with 29 beds but only one doctor, 9 nurses and a pharmacist, and one ambulance. A vaccination programme operates in the area, but there is no birth control project. AIDS is acknowledged in Tchad but rarely mentioned, and Sutton saw a large street notice in French in Bol on the subject.

The usual tropical diseases occur on the lake, malaria being the most virulent and widespread. Bilharzia is said to have increased due to the lake's current shallowness providing a greater concentration of vector snails in the shallow water. *Cholera* and *meningitis* epidemics appear also from time to time as previously and *leprosy* is endemic.

The Mission Aviation Fellowship still provides a vitally important medical lifeline from its base at N'Djamena airport around the Tchadian towns and some villages, but no longer uses amphibious planes due to the reduced open water. It is also somewhat limited by the shortage of functional airstrips.

### 6. Wildlife.

Wildlife changes have occurred too, particularly affecting the larger mammals which seem to have decreased severely in numbers. Hippos still occur in a national park on the east side of the Chari river near in its lower reaches, but the 'Lake Chad Game Sanctuary' on the Nigerian side, south of Baga appears to have been more of a map entry than an effectively managed protective area for wildlife. Otters and sitatunga are still present in some remote and less accessible areas. Crocodiles are now rarely seen, or caught in fishing nets.

Probably the most important wildlife on the lake today, apart from the fish (now reduced in availability) are the birds. A very useful description is that of Dr H. Fry[53].

He mentions 330 species recorded within a 25km radius of Malamfatori about 90 of which were palaearctic wintering regularly at Lake Chad, or passage migrants wintering further south. That however was when the lake was in its 'medium' phase. Sadly the WWF[54] reports this year a diminution in bird populations and mentions the decline of larger species such as the black crowned crane. I have not found recent statistics on pelicans or fish eagles, but red-billed queleas still seem to survive in millions, doing extensive damage to crops on all sides of the lake.

### 7. Visitors and tourists to Lake Chad.

On my 1955 and 1969/70 expeditions, I saw the lake as this majestic, if shallow, freshwater inland sea with a mystery and beauty all its own. The wealth of invertebrates, fish, birds, amphibians, reptiles, and mammals (from elephants to gerbils), was breath-taking. And all the plants: from the papyrus beds and floating islands, and from the waving *Phragmites* and *Vossia* grasses to the water lilies and convolvulus mats, captivated me. But above all else situated within that enormous, expansive, hot, dry Sahara desert of sand dunes, there were the lake people, the Yedina, so uniquely part of this 'desert of water' with their papyrus *kadais,* 'mobile homes', and their marvellous *kourri* cattle. It is very hard for tourists and visitors to the today's diminished lake

---

\* Worldwide Fund For Nature

to capture that incomparable sense of experiencing a miracle, not a mirage.[55]

## 8. The question: a future lake or a future mirage?

Who can predict the future of any large African wetland? Over the past approximately 55,000 years the great freshwater inland sea, referred to by geographers today as 'Palaeo-chad', situated originally across a huge part of the Sahara desert, *has persistently diminished* in size following a course of expansion and contraction according to climatic changes. I have outlined its supposed history in Chapter 4 of this book.

Then, some time between about 12,000 and 7000 years ago, it 'stuck' for a while at around 320m AMSL in a huge basin marked by strand lines visible in places today as ridges. This is the basin that sets a useful standard for discussion, research and politics, variously called 'Mega-Chad', 'The Conventional Chad Basin' or 'The 320m Lake Chad'.

After that it began a progressive decrease in average fluctuations until it became 'lagoonal' at just below 287m AMSL some 7000 to 500 years ago. From then on it declined further until it assumed its 19th and 20th century averages, including its striking fall in 1908 and recovery to high levels in the 1950s and 60s. So we see it has steadily shown an overall average decline in size ever since it began! *The cause:* **climatic,** *not tectonic, change.*

*Lake Chad versus The Sahara Desert*

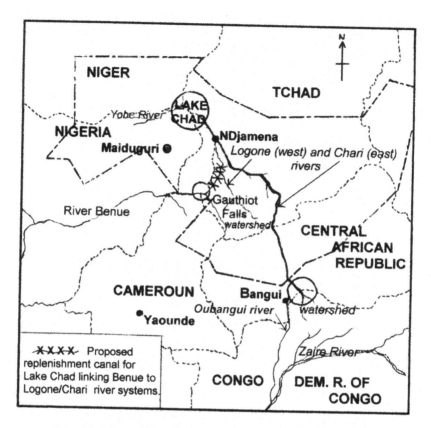

Map 15. Sketch map showing approximate position of proposed replenishment canal for Lake Chad: Lake Chad Basin Commission's 'Oubangui-Lake Chad inter-basin water transfer project'.
(Circles indicate key project sites; international boundaries.........; limit of conventional Chad basin--- - --- )

Humans could not have driven this state of affairs! There were humans around long ago following the lake on its flood plains, hunting, and maybe fishing, in the richly vegetated lagoons and forests, adapting just as today's Yedina, Hausa, Kanuri-Kanembu, Shuwa, Fulani and other tribes are doing.

It is true that they were not as numerous per sq.km as today's population, and they had not got the means to develop water management systems on a large scale. There were no big cities and adjacent nations crying out for more and more lake produce. Travel was slow - no 4-wheel-drive lorries, powered launches or aeroplanes - and life was hard: no hospitals at all, no modern medicines, no radio, e-mails or satellite telephones....But they were the forebears of the modern lake dwellers, and they survived.

The real question is: are we humans of today accelerating the effect of climatic power by our interference within the Mega-Chad catchment basin as a whole in such a manner as to hasten the final demise of modern Lake Chad itself? I believe that the effects of the irrigation systems immediately in and around modern Lake Chad are trivial (in hydrological reality) compared with the damage already done by humans to the natural protective vegetation cover of the watersheds and water courses of the Chari-Logone and Yobe rivers. Perhaps a first step is really to try to redress that damage in some way.

However 'climate change' may not primarily mean drought and desiccation, but sometimes, on the contrary, a period of torrents and floods. If that occurs, then Lake Chad may indeed return again to its 'Great Chad' levels or even to Mega-Chad levels. Climatologists seem unsure about the effects of global warming on specific areas! How right Sarch and Birkett[56] are to call for a flood warning system to help the lake dwellers to anticipate such an eventuality as major flooding. Noah warned of a flood in his day, but only 8 people believed him and survived!

We may make physical efforts to restore the lake artificially by building replenishment canals, for example, from the Benue River, but is this robbing somewhere else *(See Map 15)?*

- Can this be adequate, sustainable and realistic?
- When the vegetation protecting watersheds is decimated; when water is dammed in the watercourses and then diverted elsewhere; and when the rains fail, what water resources are left to replenish a rapidly evaporating shallow lake?
- When the sand- and diatomite- laden *Harmattan* wind dumps its burden into the lake and hastens evaporation and evapo-transpiration through plants as it blows across the lake surface, is it surprising that the hungry mobile desert dunes of

the Sahara are once again poised to take over as they did long ago across the ergs and in the Bodéle depression?

*Personally, I do not think that we humans, even with amazing future technology, can do much to turn things around to any significant degree specifically for Lake Chad unless God, the Creator-Sustainer of the universe, provides another significant climate change and actually <u>intends</u> this lake to refill.*

# References

1. (page 6) FALCONER, J.D. (1911), On horseback through Nigeria Fisher Unwin, London)
2. (page 6) BARTH, H. (1890), Travels and discoveries and North and Central Africa (Ward, Lock & Co., London and Manchester)
   – (1857), Travels and discoveries in North and Central Africa (Longmans, London), 5 vols
3. (page 7) DENHAM, D. (Major) and (Capt) CLAPPERTON (1826), Travels and Discoveries in Northern and Central Africa (London, Murray, 2 vols)
4. (page 8) ALEXANDER, BOYD (1907), From the Niger to the Nile (Edward Arnold, London)
5. (page 32) FALCONER, J.D. (1911), Geology and geography in Northern Nigeria (Macmillan, London)
   – the Bornu, Sahara, and Sudan (Murray, London)
6. (page 88) OLIVRY et al, (1996) Hydrologie du Lac Tchad, ORSTOM éditions, Paris (With a useful bibliography).
7. (page 108) PIAS, J. (1958), Transgressions et regressions du Tchad à la fin d'ère Tertiare et du Quaternaire (C.r.AC.SC, Paris), 246:800-(1967), Quatre deltas successifs du Chari au Quaternaire (Republique du Tchad et du Cameroun) (C.r.AC.SC, Paris), Series D, 264:2357-60
8. (page 110) PULLAN, R.A.(1965) The recent geomorphological evolution of the south-central part of the Chad Basin, Samaru res.Bull.58 (A.B.U. Zaria. Northern Nigeria)
9. (page 111) FALCONER (1911) - op cit-
10. (page 112) MIGEOD,F.M.H.(1924) Through Nigeria to Lake Chad.Geol. notes to Lake Chad ch 13 (Heath Cranto, London)
11. (page 113) PULLAN. op.cit.
12. (page 116), TILHO), J. (1910-1914), Documents scientifiques de la Mission Tilho, 1906- Paris (Imprimerie nationale), Vol. I, 1910, Vol II, 1911, Vol. III, 1914.
13. (page 116) GROVE AND WARREN (1968), 'Quaternary landforms and climate on the south side of the Sahara', Geographical Journal 134, pp.194-208.

## References

14 (page 120) du PREEZ and BARBER, W. (1965), Pressure water in the Chad Formation of Bornu and Dikwa emirates, North-eastern Nigeria, Geological Survey of Nigeria, Bulletin 35, (Federal Government of Nigeria

15 (page 123) Miller et al (1965) Availability of groundwater in the Chad basin of Bornu and Dikwa emirates, Northern Nigeria, U.S.Survey with Geol. Survey of Nigeria (U.S.Agency for International Development).

16 (page 144) HEPPER,F.N. (1970) plant life of sandbanks and papyrus swamps. Geog.mag.XLII,8:557-82.

17 (page 144) HOPSON,A.J. (1968) Notes on the aquatic flora of the northern basin of Lake chad. Pers.Comm.

18 (page144) ROBINSON, A. and P. (1967), Preliminary observations on the plankton of the northern basin of Lake Chad, Annual Report 1966-7, Fed. Fish. Services, Lake Chad Res. Station, Malamfatori, Nigeria

19 (page 144) LEONARD, J. (1969), Apercu sur la vegetation du Lac Tchad, compliment au Chapter 1 de la Monographie Hydrologique Lac Tchad O.R.S.T.O.M., Paris) 20 (page 155)

20 (page 147) GRAS, R., ILTIS, A., and LEVEQUE-DUWAT, S. (1967), Le plankton du Bas-Chari et de la partie est du Lac Tchad (O.R.S.T.O.M., Paris)

21 (page 147) HUTCHINSON and DALZIEL, (1954). Flora of tropical Africa Vol.I, 2nd edn ed Keay (1954) (Crown Agents, London)

22 (page 155) ROBINSON,A and P. (1970), Seasonal distribution of zooplankton in the northern basin of Lake Chad, Journal of Zoology (London), 163:25-6121

23 (page 156) HOPSON, A.J. (1967),Op.Cit.

24 (page 158) DUPONT and LEVEQUE (1967) Biomasses en mollusques et nature des fonds dans la zone est du Lac Tchad (ORSTOM) Paris.

25 (page 158) DEJOUX, C. (1967), Contribution à l'études des insectes aquatiques du Tchad (O.R.S.T.O.M., Paris

26 (page 160) HOLDEN And REED (1972) West African freshwater fish, Longman Group Ltd, London.

27 (page 169) BANNERMAN, D.A. (1952-3), Birds of West and Equatorial Africa, Vols. I and II (Oliver & Boyd, London).

ASH, J.S., FERGUSON-LEES, I.J., and FRY, C.H. (1967), Expedition to Lake Chad, Northern Nigeria, 1967, Preliminary Report, Ibis 109, 478-86

28 (page 178) Burgoin, P. (1955), Animaux de chasse d'Afrique (La Toison d'or, Paris)
ROSEVEAR, D.R. (1953), Checklist and atlas of Nigerian Mammals (Government Printer, Lagos)
JEANIN, A. (1936), Les mammiferes sauvages du Cameroun (Paul Lechevalier, Paris)

29 (page 179) DENHAM, D., op.cit.
30 (page 179) ALEXANDER, Boyd., op. cit.
31 (page 205) BELL, W.D.M.(1960).Bell of Africa.
32 (page 208) SIKES, S.K. (1971) The African Elephant.. Weidenfeld and Nicholson. London.
33 (page 215) COHEN,R (1967) The Kanuri of Bornu (Holt, Reinhart & Winston) New York.
34 (page 215) KOELLE (1854), Grammar of the Bornu or Kanuri Language (C.M.S., London)
(page 215) TEMPLE, O. (1922), Notes on the tribes, provinces, emirates and states of the Northern provinces of Nigeria (Frank Cass, London)
(page 215) URVOY,(1949), Histoire de l'Empire du Bornu, Mem. De l'Inst. Française d'Afrique Noire, Vol 7 (Lib. Larousse, Paris)
(page 215) LENFANT, E.A. (1905), La grande route du Tchad

35 (page 215) BENTON, P.A. (1913), The sultanate of Bornu, translated from German of Dr A. Schultze, with additions and appendices (O.U.P, London) page 157
36 (page 215) FALCONER, J.D., On horseback through Nigeria
37 (page 215) MURDOCK, G.P. (1959), Africa: its peoples and their culture (McGraw-Hill, New York)
CHEVALIER, A. (1908), Mission Chari-lac Tchad, 1902-4
CORNEVIN, R. (1960), Histoire des peuples de l'Afrique
FAGE, J.E. (1962), An introduction to the history of West Africa (C.U.P., Cambridge)
BOVILL,E. (1933), Caravans of the old Sahara (O.U.P., London)

*References*

HOGBEN, S.J. (1930), The Muhammedan emirates of Nigeria
TALBOT, P.A. (1911), 'The Buduma of Lake Chad', Journal of the Royal Anthropological Institute (London) MEEK, C.K. (1925) The northern tribes of Nigeria (O.U.P., London)

38   (page 216). DENHAM, D., op.cit.
39   (page 232).BARTH, H.,op.cit
40   (page 233) FALCONER. op.cit.
41   (page 237) DENHAM). op.cit.
42   (page 237) BARTH. op.cit.
43   (page 242) TEMPLE. op.cit.
44   (page 318) BARRINGTON.(1973),Rev.Bro Leo, Personal communication.
45   (page 339) SIKES,S.K. (1972) Lake Chad, Eyre Methuen, London.
46   (page 339) SIKES. op.cit.
47   (page 339) OLIVRY et al. op.cit.
48   (page 340) SUTTON,J.(2002). Personal Communication.
49   (page 342) WWF (2002).www.worldwildlife.org
50   (page 346) SARCH,T. and BIRKETT, C.Fishing and farming at Lake Chad: responses to lake-level fluctuations. Geog. J, vol.166, no.2,pp 156-172
51   (page 348) .CHILVERS.A and SAMA'ILA, Rev.M (2002). Personal communication.
52   (page 349) SUTTON op.cit.
53   (page 352)  FRY, H (1971) Lake Chad: retrospect and prospect. The Nigerian Field Vol.XXXVI. No 3. pp 100-114.
54   (page 353) WWF (2002). www.worldwildlife.org
55   (page 354) de VILLIERS, M. and HIRTLE, S (11997) Into Africa  pp 270-273. Phoenix, london.
56   (page 357) SARCH AND BIRKETT. op.cit.